Geoheritage, Geoparks and Geotourism

Conservation and Management Series

Series Editors

Wolfgang Eder, GeoCentre-Geobiology, University of Göttingen, Göttingen, Niedersachsen, Germany

Peter T. Bobrowsky, Geological Survey of Canada, Sidney, BC, Canada

Jesús Martínez-Frías, CSIC-Universidad Complutense de Madrid, Instituto de Geociencias, Madrid, Spain

Spectacular geo-morphological landscapes and regions with special geological features or mining sites are becoming increasingly recognized as critical areas to protect and conserve for the unique geoscientific aspects they represent and as places to enjoy and learn about the science and history of our planet. More and more national and international stakeholders are engaged in projects related to "Geoheritage", "Geo-conservation", "Geoparks" and "Geo-tourism"; and are positively influencing the general perception of modern Earth Sciences. Most notably, "Geoparks" have proven to be excellent tools to educate the public about Earth Sciences; and they are also important areas for recreation and significant sustainable economic development through geotourism. In order to develop further the understanding of Earth Sciences in general and to elucidate the importance of Earth Sciences for Society, the "Geoheritage, Geoparks and Geotourism Conservation and Management Series" has been launched together with its sister "GeoGuides" series. Projects developed in partnership with UNESCO, World Heritage and Global Geoparks Networks, IUGS and IGU, as well as with the 'Earth Science Matters' Foundation will be considered for publication. This series aims to provide a place for in-depth presentations of developmental and management issues related to Geoheritage and Geotourism in existing and potential Geoparks. Individually authored monographs as well as edited volumes and conference proceedings are welcome; and this book series is considered to be complementary to the Springer-Journal "Geoheritage".

Javier Dóniz-Páez • Nemesio M. Pérez

Editors

El Hierro Island Global Geopark

Diversity of Volcanic Heritage for Geotourism

 Springer

Editors
Javier Dóniz-Páez
Departamento de Geografía e Historia
Universidad de La Laguna
San Cristóbal de La Laguna, Spain

Nemesio M. Pérez
Instituto Volcanológico de Canarias
(INVOLCAN)
San Cristóbal de La Laguna, Spain

ISSN 2363-765X   ISSN 2363-7668   (electronic)
Geoheritage, Geoparks and Geotourism
Conservation and Management Series
ISBN 978-3-031-07291-8   ISBN 978-3-031-07289-5   (eBook)
https://doi.org/10.1007/978-3-031-07289-5

This Springer imprint is published by the registered company Springer Nature Switzerland AG
The registered company address is: Gewerbestrasse 11, 6330 Cham, Switzerland

Preface

El Hierro is the smallest and geologically the most recent island of the Canaries, and it registered the last submarine eruption in Spain during the years 2011 and 2012. It is an oceanic, subtropical and volcanic island which has a low population pressure and as a result has barely modified its original volcanic and non-volcanic landscapes. All these aspects allow the existence of a great geodiversity of volcanic morphologies (cinder cones, lava flows, lava deltas, lava tubes, hornitos, tumuli), erosion processes (landslides, ravines and cliffs) and sedimentary processes (beaches, dunes, alluvial and colluvial deposits) which have been geoconserved. In addition, more than 52% of its territory are natural protected areas by the Canary Law of Natural Spaces, for example a Biosphere Reserve and a UNESCO Global Geopark, and aiming also to be energetically self-sufficient. Moreover, El Hierro Island contains a singularity of rural landscapes associated with its volcanic origin, its subtropical latitude, its scarce waters and low population Tourism in the island is a sustainable activity, and its main attractions are diving and hiking through the different volcanic landscapes of the island. All these aspects contribute to the geographical diversity of El Hierro in reference to its volcanic (heritage) and rich cultural heritage. Because of this, the informative nature of this book becomes necessary, written in a simple but scientific language, allowing this way the main readers to be scientists specialized in geotourism, active leisure entrepreneurs and the general public interested in volcanic geoheritage and geotourism.

The chapters included in this book provide a general but also detailed overview of the main aspects that characterize El Hierro, its Global Unesco Geopark and the integration of natural volcanic and non-volcanic geoheritage with its society along with the history of the island and the heritage generated. The book is structured into four parts. The first part is an introduction about the importance of geoheritage and its relationship with other concepts such as geodiversity, geoconservation, geoculture and geoparks. The second part is dedicated to the geography and geology context of the island. The third concerns the diversity of geographical natural and cultural heritage. And finally, the fourth part is dedicated to geotourism and the main products in El Hierro Global Geopark promoting its sustainable use. Thus, through the different chapters of this book we will learn about the main values associated with the geology, landscapes, habitats, history and culture of El Hierro.

Chapter "Volcanic Geoheritage in the Light of Volcano Geology". It introduces the main concept associated with geoheritage, geodiversity and geoconservation in volcanic landscapes with special reference to the highlight of geology perspective. Moreover, the chapter revises, in detail, the natural and cultural heritage present in volcanic landforms and processes.

Chapter "Volcanology of Recent Oceanic Active Island". It describes the main volcanological features of El Hierro focusing on the origin of the island and its evolution and identifies its main geological edifices. The chapter offers the reader an overview of the geology of El Hierro and its current scientific knowledge.

Chapter "Volcanic Geomorphology in El Hierro Global Geopark". It helps to understand the volcanic geomorphology of El Hierro and its volcanic and non-volcanic diverse forms and processes. The chapter explains, in detail, the physiography features, the erosive and accumulative landforms and finally the monogenetic mafic volcanism present in the island.

Chapter "Geoheritage Inventory of the El Hierro UNESCO Global Geopark". It shows the vegetation associated with the landscapes of El Hierro and its relationship with the volcanic relief, with special attention to the evolution of the vegetation in volcanic rifts of El Hierro Geopark. It also reviews the great diversity of species and habitats.

Chapter "The Vegetation Landscapes of a Oceanic Recent Volcanic Island". It describes the main volcanic geomorphosites detected in El Hierro Global UNESCO Geopark and their promotion of the itineraries for volcano tourism in this island. In this chapter, one of the georoutes in the island is selected in order to show the diversity of volcanic and non-volcanic geoheritage, spectacular vegetation landscapes and a rich cultural heritage.

Chapter "Human Occupation of a Small Volcanic Island". It describes the identification, selection and characterization of the main geosites present in El Hierro Global UNESCO Geopark. These geological and geomorphological sites show its geoheritage which are very important to the new geotouristic products in the island.

Chapter "Rural Landscapes in an Oceanic Volcanic Island". It describes the main geomorphosites of El Hierro based on the diversity of its volcanic and non-volcanic geoheritage and cultural heritage and proposes a georoute with volcano tourist interest in the Orchilla geozone, where there is an important place of the island with a diverse and rich natural and cultural heritage associated with the mafic volcanism.

Chapter "Geomorphosites of El Hierro UNESCO Global Geopark (Canary Islands, Spain): Promotion of Georoutes for Volcanic Tourism". It is devoted to the geographical biodiversity of the geopark and the great variety of birds and their habitats. And then, a trail network of paths and viewpoints of El Hierro is proposed which constitutes the basic infrastructure for ornithological tourism or birdwatching under the sustainable uses' principles.

Chapter "Birdwatching as a New Tourist Activity in El Hierro Geopark". It shows the cultural seascapes in the marine reserve of "Sea of Calms" and La Restinga coast. This geographical space is the main tourist destination of El Hierro Geopark, and the main activities are associated with scuba diving. The biodiversity of marine life and the cultural heritage of the fishers are two principal characteristics of this landscape.

Chapter "Cultural Seascapes in the 'Sea of Calms' and La Restinga Coast" reviews the submarine eruption which occurred in 2011–2012 in the Mar de las Calmas Marine Reserve. This is the first eruption occurred in El Hierro in the historical period. The chapter is divided into two sections. In the first one, the eruption is analysed, and the second shows the geotouristic interest of this eruption. Lastly, we will remember when this eruption finished since the local administration started the project for El Hierro as a Geopark in 2012.

Finally, Chapter "Submarine Eruption of El Hierro, Geotourism and Geoparks" reviews the submarine eruption which occurred in 2011–2012 in the Mar de Las Calmas marine reserve. This is the first eruption occurred in El Hierro in the historical period. The chapter is divided in two sections. In the first one the eruption is analysed and the second shows the geotouristic interest of this eruption. Lastly, we will remember when this eruption finished since the local administration started the project for El Hierro as a Geopark in 2012.

This book was supported by project "VOLTURMAC, Fortalecimiento del volcano turismo en la Macaronesia (MAC2/4.6c/298)", and is co-financed by the Cooperation Program INTERREG V-A Spain-Portugal MAC (Madeira-Azores-Canarias) 2014–2020, http://volturmac.com/.

Tenerife, Spain   Javier Dóniz-Páez
                                                                          Nemesio M. Pérez

Contents

Volcanic Geoheritage in the Light of Volcano Geology

Károly Németh

Abstract

Volcanic geoheritage relates to the geological features of a region that are associated with the formation of a volcanic terrain in diverse geoenvironmental conditions. These features include the volcanic processes, volcanic landforms and/or the eruptive products of volcanism that form the geological architecture of that region. Volcanic geoheritage is expressed through the landscape and how it forms and evolves through volcanic processes on various spatio-temporal scales. In this sense it is directly linked to the processes of how magma released, transported to the surface and fragmented, the styles of eruption and accumulation of the eruptive products. Volcanic geoheritage is directly linked to the natural processes that generated them. Geocultural aspects are treated separately through volcanic geosite identification and their valorization stages. Identification of volcanic geosites, based on various valorization techniques, have been applied successfully in the past decades to many geological heritage elements. Volcanism directly impacts societal, cultural, and traditional development of communities, hence the "*living with volcanoes*" concept and indigenous aspects and knowledge about volcanism can and should play important roles in these valorization methods through co-development, transdisciplinary approaches by including interconnected scientists in discussions with local communities. Elements of volcanism and volcanic geoheritage benefit of the geoculture of society so volcanic geoheritage sites are ideal locations for community geoeducation where resilience toward volcanic hazard could be explored and applied more effectively than it is done today. Geoparks within volcanic terrains or volcanism-influenced regions should be the flagship conservation, education and tourism sites for this message. Volcanism can be an integral part of processes operating in sedimentary basins. Here volcanic eruptive products and volcanic processes contribute to the sediment fill and geological features that characterize the geoheritage of that region.

Keywords

Geoheritage • Geodiversity • Geoconservation • Geoeducation • Volcanic facies • Eruption style • Explosive • Effusive • Pyroclastic • Volcano geology • Monogenetic • Polygenetic • Stratovolcano • Caldera • Submarine volcanism

1 Introduction

Volcanic eruptions are frequently the subject of global and local media attention because volcanism fascinates people, even in areas not hosting active volcanoes. In fact, volcanic events generate more interest from people than any other geological processes (Erfurt-Cooper 2011; Erfurt-Cooper 2014) (Fig. 1). This behavior has been identified as one of the main driving forces behind volcano tourism, a special type of geotourism associated with adventure tourism (Erfurt 2018). Volcanic geology has been incorporated into methods for evaluating the geoheritage values of volcanic terrains especially from the perspective of UNESCO World Heritage site nominations. The main international body that stands behind the UNESCO World Heritage site selections mostly by providing advice and recommendation of the scientific value of the proposed sites, The International Union for Conservation of Nature (IUCN), has published two thematic

K. Németh (✉)
School of Agriculture and Environment, Massey University, Palmerston North, New Zealand
e-mail: K.Nemeth@massey.ac.nz

K. Németh
Geoconservation Trust Aotearoa Pacific, Ōpōtiki, New Zealand

Institute of Earth Physics and Space Science, Sopron, Hungary

Fig. 1 Volcanic eruption during the 2021 Geldingadalir, Iceland eruptive events the full array of a typical mildly explosive, dominantly effusive basaltic volcanic eruption can be observed (**a**) that fascinate the visitors (**b**). Photo by *Gisli Gislason*

studies on volcanoes (Wood 2009; Casadevall et al. 2019). The first issue, published in 2009 (Wood 2009), outlines the significance of volcano science and volcanic landforms in the selection criteria for granting UNESCO World heritage site status to a volcanic terrain. This report, however, lacks a practical, systematic comparative study that nominating bodies could readily deploy. Hence where sites proposed for UNESCO World Heritage status had strong association with volcanism it became apparent that further study was warranted. A new report, released in 2019 as the World Heritage Volcanoes document (Casadevall et al. 2019), recommends classification methods, knowledge gap analysis and some recommendations about how future sites should be accepted for listing. While this report, and the stronger involvement of IUCN in this process, with the aid of geoscientists with expertise on volcano geology is certainly a major step forward, it is still very general and lacks definitive guidelines.

Volcanic geoheritage is currently used in a very broad sense, essentially to any volcanic terrain, feature, processes, deposit or eruption that are in some way unique to or associated with some geocultural perspective (Nemeth et al. 2017). The historic eruption record, oral traditions from indigenous cultures or strong geocultural links are used to define the volcanic geoheritage. Volcanic geoheritage is commonly viewed as an attribute serving geotourism or geoeducation purposes. While there is no doubt that volcanism is a key element of many geotourism projects, development of geopark models and heavily linked to its geoeducation potential, volcanic geoheritage is somehow a more broader concept and it should be viewed through our current knowledge on how volcanoes work or evolve (Fig. 2a), their role in creating or modifying landscapes

(Fig. 2b, c), and how they vanish over time (Fig. 2d). In this chapter we provide a working approach to view volcanic geoheritage as a universal and absolute value of geoheritage that also provides a scientifically established background of qualitative and quantitative geodiversity estimates of volcanic terrains or volcanism-influenced sedimentary basins. The proposed approach provides a firm foundation of how volcanic geoheritage can be utilized in geoconservation strategies or for geotourism purposes. Here a proposed approach is outlined that provides a non-biased, geologically validated approach to express the attributes of volcanic geoheritage. Later, we discuss the geocultural aspects including indigenous cosmovisions on volcanism that can act as a driving mechanism to valorize geoheritage values to identify, locate and map volcanic geoheritage sites.

2 Geoheritage—Geodiversity— Geoconservation from Volcano Science Perspective

There is general confusion and convoluted usage of terms and methods used to define volcanic geoheritage. The same issue exists in how we treat and define geoheritage in general. Current systematic research, based on study of the published scientific data shows that a largely inhomogeneous approach exists to define geoheritage and consensus has not been reached yet (Nemeth et al. 2021a, 2021). In many cases geoheritage, geoheritage site and geodiversity are mixed terms that are inconsistently applied with few if any synonym terms. "*Geoheritage*" is a generic but descriptive term applied to sites or areas of geologic features

Fig. 2 **a** Eruptions styles control the overall architecture and even the volcanic landform characteristics as well as they reflected in the deposit characteristics in meso and micro scales. In the Motukorea/Browns Island in the quaternary Auckland volcanic field a typical basal phreatomagmatic tuff ring section is capped by magmatic explosive eruptions that generated scoriaceous capping units. **b** Most of the volcanic eruptions have short to medium term (e.g., thousands of years to maybe few millions of years) landscape-modifying effects. This can be observed from Mount St. Helens that after the 1980 eruption deposited extensive volcaniclastic fans in the ring plain. Since the eruption, the newly accumulated volcaniclastic fan gradually incised by local stream network and remobilized and redeposited significant part of it providing and advancing volcaniclastic fan in the distal regions of

the volcano. **c** In exceptional cases volcanic eruption are significant landscape producing events that shape the landscape dramatically as the case in most of the large ignimbrite flare up events in the Earth History such as the Pliocene ignimbrite plateaus at the Andean Central Volcanic Zone in Chile. The Pliocene ignimbrite sheets cover the region over several hundreds of metres thick ignimbrite successions providing light pink tone to the landscape and acting as a base on the Pleistocene and Holocene intermediate (andesite, dacite) polygenetic compound, composite and stratovolcanoes grown in the last hundreds of thousands of years. **d** Erosion can alter volcanic landscape and produce visually and aesthetically unique landscape elements such as the dissected and exhumed core of Miocene stratovolcanoes proximal sections at the Fletcher Bay of the Coromandel peninsula, New Zealand

with significant scientific, educational, cultural, and/or aesthetic value (Brilha 2018b; Macadam 2018). According to this widely adapted definition geoheritage is widely accepted as a site-specific definition that can be defined by various valorization techniques including its scientific, educational, cultural and aesthetic values (Brilha 2016, 2018a). The value is heavily influenced by the societal and cultural activities associated with the site. However, if we wish to express, for comparative reasons, the geoheritage values of a specific site's essential to do so in its geological and geomorphological context independent of human society and culture. Geoheritage is something that reflects the heritage elements of Earth History and the processes that generate the geological and geomorphological features (Fig. 3). Geoheritage in this perspective treats Earth's processes and history as the controlling factors regardless of Recent interactions with human society. Geoheritage sites are those which has been

identified and studied to generate inventories and comparative studies expressing the relative values of those locations (Brilha 2016) (Fig. 4). In this context scope and scale of the features are important. Within the framework of valorization, the various methods applied are generally linked and heavily dependent on the recognized scope of the site. The most common approach is for geotourism where the valorization shows strong linkages to the cultural activities associated with the identified geosite. Hence strong arguments can be made to include geocultural or even indigenous elements in the geoheritage element description. The perspective proposed here aims to avoid this confusion by putting geoheritage as the absolute value on which to base our current and best scientific view of a geological or geomorphological entity (Fig. 3). To create a workable framework to identify the geoheritage elements a systematic overview looking at the geological and geomorphological features from a topological aspect (e.g., descriptive sense) and the processes operating at the time of their creation (Fig. 5). The processes then can be interpretative as they are commonly inferred from some measurable and/or observable parameters in the geological or geomorphological record. Intuitively, this method should be independent of the way we gain information to solve the problems, e.g., how the arguments and assessment probe the observed geoheritage element. To put it simply, the geoheritage value of a geological feature should be entirely dependent on the current scientific understanding of the processes that created the feature regardless of the pathway followed to reach that knowledge.

Geodiversity, in contrast, is something that expresses the diverse nature of the geoheritage elements from the perspective of the processes that formed the feature (Gray 2018a, b; Zwoliński et al. 2018). Geodiversity is commonly defined as the variety/natural range/range/diversity of the non-living (abiotic) environment. This assessment of a feature is intended to be all-embracing encompassing 'geological nature/geological features/the geological environment'. This definition holds two logical directions, (1) all geological and geomorphological elements be included and (2) determine how this variety of features can be expressed. There are those who propose geodiversity should be treated in a similar way to biodiversity, as an expression of number/density of features (Fig. 6). This is a promising and interesting idea, but very difficult to develop and test. While the idea is commendable and would expand the current application of biodiversity to include the abiotic aspects there are some significant issues that need to be explored further. Namely, biodiversity calculations are almost exclusively based on a selection of species and look at their population density within various regions. The selection of a species is a kind of "*singularity-defined*" approach as the

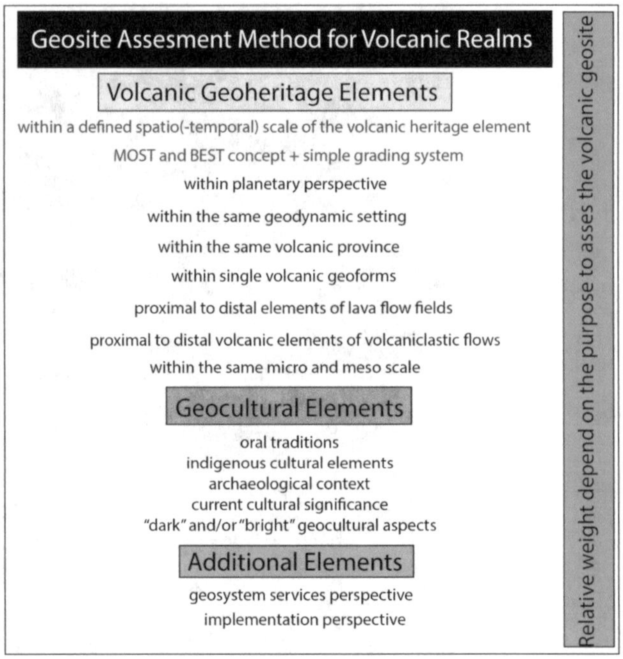

Fig. 4 Geosite valorization aspects in volcanic context as a model that needs to be specifically designed for volcanic terrains. The methods essentially based on the main versus additional values that are pre-set for the scope of the valorization such as in most of the cases serving geotourism. The methods in volcanic terrains should reflect the identified volcanic geoheritage elements as a key framework along the search for the "most and best" site should be identified

definition of species—especially in comparison to the scale of the observations (and measurements) are well defined, and easy to adapt. In a simple way, we know exactly how to identify the specific species and we can develop methods about how to measure their appearance in a specific spatial dimension. The definition of a species, or even higher

Fig. 3 Conceptional framework to outline the interlinkages of geoheritage elements—geoheritage sites and geodiversity and their connection to geotourism and geoeducation specifically expressed in a volcanic geological context

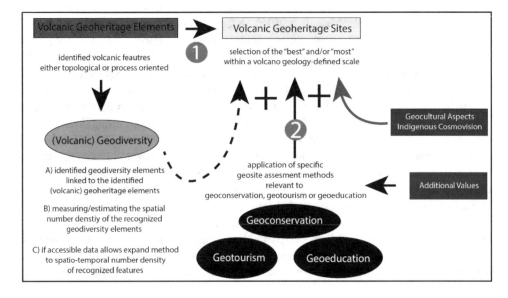

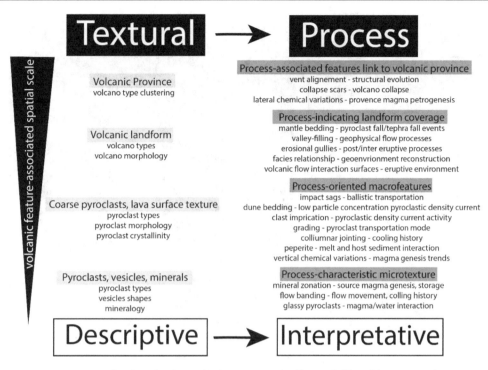

Fig. 5 Textural versus process associated volcanic geoheritage elements. Textural elements can be linked to descriptive features, for instance elements that can be expressed in some geometrical scales (e.g., landforms, fields, edifice etc.), while process elements are those that show strong link to a specific key volcanic geology process

biological orders are clear and tested over many decades, hence no, or little, ambiguity exists within the counting methods. The measurable biotic elements (e.g., a species of animal) that move across the spatial region in question, are generally small compared to the size of territory and the biodiversity being defined. However, we have no guidance or even recommendation on the best way to define the geological or geomorphological "unit(s)" we wish to measure in similar context. The current recommendations that geodiversity elements be as inclusive as possible can create confusion as various elements could be treated with equal weight in the estimation methods (Coratza et al. 2018; Zwoliński et al. 2018). Often geological and geomorphological elements are sizable with dimensions similar to the study area where the geodiversity is being assessed. To assess the spectrum of landscapes, landforms and geological/geomorphological processes if a region we need to define the measurable geological features to be included in the geodiversity calculation. Geodiversity estimation produced a great number of outputs recently commonly utilizing advanced technologies as spatial statistic or GIS applications (Benito-Calvo et al. 2009; Bradbury 2014; Argyriou et al. 2016; Araujo and Pereira 2018; Betard and Peulvast 2019; Albani et al. 2020). Probably the best approach is to utilize the available geological maps at various mapping scales; these provide the raw data pertaining to the geological elements of the region. The geological maps

provide lithology-based information which are relatively easy to define, identify, or reproduce even by end-users not deeply involved in geological research.

On the basis of the geological maps, the measurable elements can be expanded to include processes associated with the lithological entities. In addition, geological maps contain data about the structural elements of the region. These are another measurable variable.

Overall, the geoheritage, geoheritage site and geodiversity are three unique features interconnected by a conceptual framework that can be applied to define a region (Coratza et al. 2018). The scope and scale are important aspects when defining these three parameters. The scope will include either descriptive or process-related aspects of the above elements but also can be linked to the purpose of the research (e.g., geotoursim, geoeducation, urban planning etc. purposes). Scale is important and is commonly associated with research purpose-defined approaches and is looked at from regional, national or global aspects. This is a valid and functional approach, especially for geotourism, but probably not the best when studying geoheritage elements. Geoheritage elements should be investigated from an internal scale perspective and relate to the dimensions of the geological elements under investigation (Fig. 7). For instance, a specific sedimentary basin that produced a specific sediment deposit that lithified into sedimentary rock has natural (and measurable) spatial and temporal dimensions. Hence when

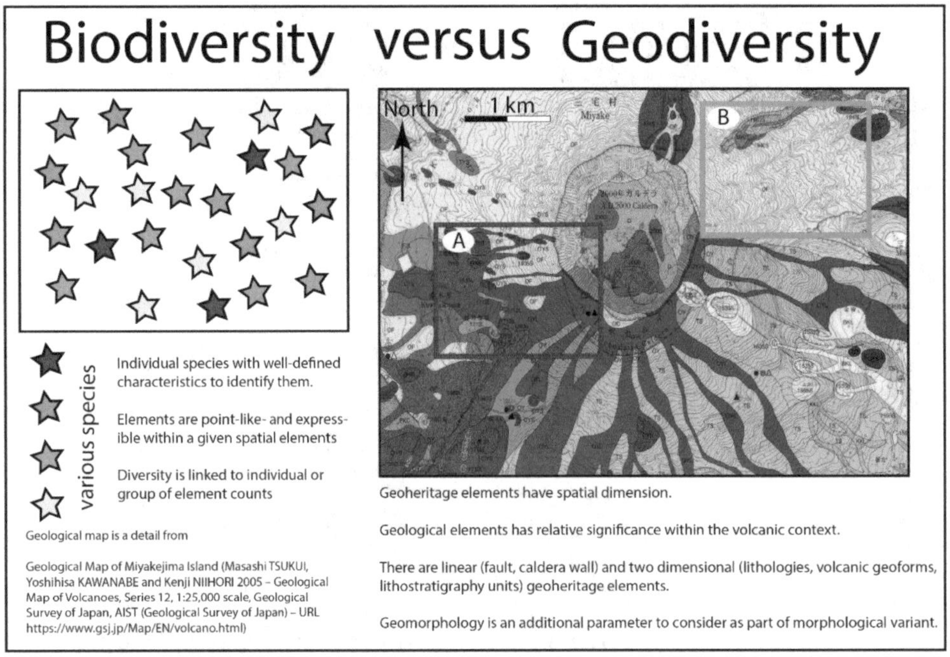

Fig. 6 Comparison of bio- and geodiversity highlighting the challenged geodiversity estimates faces with. Geological map is a detail from the Geological Map of Miyakejima Island (Masashi TSUKUI, Yoshihisa KAWANABE and Kenji NIIHORI 2005—Geological Map of Volcanoes, Series 12, 1: 25,000 scale, Geological Survey of Japan, AIST - available on https://www.gsj.jp/Map/EN/volcano.html). Note the potential variation of geodiversity based on the mapped volcanic geological elements marked on the geological map with 1–25,000 scale. "A" region has more geological elements than "B". While intuitively we estimate significantly higher geodiversity for the area "A", this needs to be evaluated as the large "orange" zone OF (Pleistocene Ofunato stage products) are complex volcanic succession. In modern volcanic settings, geological maps can be heavily biased by mapped erupted products toward younger features that is not obviously reflecting the created volcano geology diversity of the system. Such problem needs to be treated carefully as so far, no general strategy exists to deal with the timescale and volcanic architecture resolution. In general, we can assume that this problem eases toward Cenozoic volcanic terrains and geological maps can be more reliable source for geodiversity calculation in relationship with other geological entities mapped

looking at geological heritage elements that formed over time the time during which that element formed needs to be considered as well.

Overall, this theoretical approach to geoheritage should also embrace the concept of geological facies. For example, the typical geoenvironment in those features formed and the physical and chemical processes responsible for generating those geological features.

The following sections will explore how volcanic geoheritage fits to this theoretical approach and what makes volcanic geoheritage unique in respect to other geological and geomorphological phenomena.

3 Geoheritage Recognition and Value Estimates from Volcano Geology Perspective

Volcanoes are special geological features that manifest in great variety across the Earth and have throughout history have displayed diverse eruption styles, many of which are not occurring today. Volcanic eruptions can be classified according to the power (e.g., VEI—volcanic explosivity index) of the eruption and their potential to modify an existing or create a new landscape (Newhall and Self 1982; de Silva and Lindsay 2015) (Fig. 8a). The most frequent eruptions in Earth history are of moderate intensity (VEI = 1–4) producing eruptions that modify the landscape and with eruptive products that are "*locked*" into the geological record, commonly as part of sedimentary basin successions. The lifespan of such volcanic eruptions can be very short (days to weeks) up to over 100,000 years. This type of volcanism is known across the Earth surface and in world's oceans and typically produce monogenetic or polygenetic volcanoes (Fig. 8b). Monogenetic volcanoes in this context refers to the eruption that feed from a single volcanic conduit and last for only short period of time (hours to maybe years) and are clearly associated with a single batch of magma that finds its way to the surface via the same conduit. In contrast, polygenetic volcanoes are those that form and establish plumbing systems (de Silva and Lindsay 2015; Nemeth and Kereszturi 2015; Smith and Németh 2017). These are commonly fed from crustal magma storage places and can evolve over numerous eruptive episodes to build an

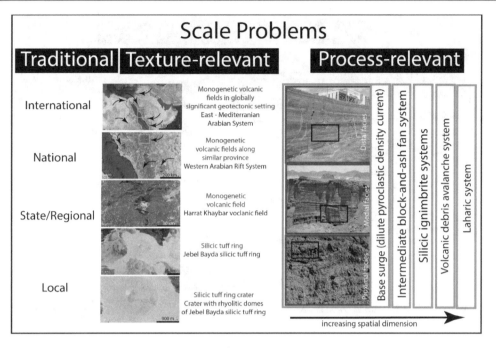

Fig. 7 Scale "problem" in a graphical expression. Conventional scale setting used in geoheritage valorizations are contrasted with volcanic process and features viewed in the scale the identified elements prescribe. The graphical representation is challenging to show process-relevant elements. The images show eruptive products highly relevant to processes in various scales following the volcanic facies concepts. The image frames refer to a typical monogenetic volcano-associated volcanic features (e.g., pyroclastic density current deposition). Green rectangles show other key geological processes associated with volcanism. Within those features various scales can be identified following the volcanic facies model

amalgamated and complex volcanic edifice such as a strato- or compound volcano. Polygenetic eruptions are only capable of some modification to the surrounding environment and their preservation potential is largely controlled by the climatic and geoenvironmental conditions. Over time, only the conduit or proximal volcanic successions are preserved, commonly forming distinct landscapes with volcanic plugs, exhumed upper conduits or completely inverted landscapes. Large, normally silicic (e.g., rhyolitic) eruptions commonly form extensive pyroclastic blankets such as ignimbrite sheets and distinct collapse features such as calderas all characteristically changing the appearance of the pre-eruptive landscape, hence these eruptions commonly referred to as landscape forming eruptions (Graettinger 2018) (Fig. 8c).

Volcanism produces spectacular geological features, however, in the last 600 million years volcanic geological elements are volumetrically well below that of sedimentary or metamorphic rock types, hence they are in general rare events despite their huge local impact, making volcanic geoenvironments a distinct geoheritage type. In addition, volcanism generally takes place in a far shorter time frame than any other geological processes thereby making their eruptive products valuable chronostratigraphic markers for understanding Earth history (Fig. 8d).

Volcanoes themselves and the associated eruptive products are diverse in their appearance. Defining the variety of volcanic features that may be associated with a volcano is a complex task, in part partially because they form much faster than other geological elements. Furthermore, their physical appearance is governed by their unique chemical and physical processes. The general magma chemistry, volatile content and the magma petrogenetic conditions all add to the complexity of the type of volcanic eruption that may result. The connection between the magma petrogenetic features and the geotectonic environment they are most likely to form in make volcanism and volcanic geoheritage the perfect avenue to link and identify key volcanic features used to define volcanic geoheritage elements. As a proxy the mapped rock types, such as basaltic to rhyolitic, could be used as distinct attributes associated with the volcanic geoheritage. Each of the mapped rock types somehow can be or should be linked to larger geotectonic processes such as convergent plate margin, intraplate, ocean island etc. setting.

Using volcano architecture through volcanic facies models (Cas and Wright 1987; Gamberi 2001) (Fig. 8f) enables the volcanic geoheritage to be defined by the petrochemical elements, and their link to the geotectonic features (De Vries et al. 2018) (Fig. 8e). Every volcano type has a point source, a vent that is linked to a conduit/magmatic plumbing system and through a crater connect to the proximal volcanic edifice. By increasing distances from the vent, the most common or typical

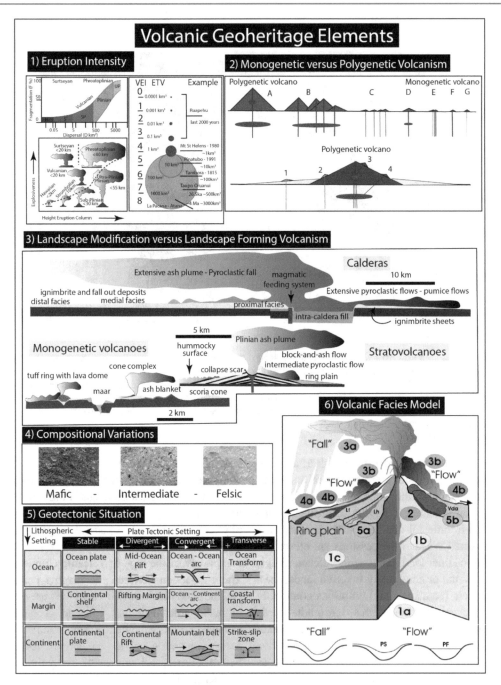

Fig. 8 Complex diagrammatic representation of key volcanic geoheritage elements. Please note, that these are just the most representative geological features that play key to define any volcanic systems. 1. Volcanic eruption intensity (size of an eruption) represented in three commonly referred cartoon where increased explosivity also means higher, more vigorous eruption plumes and higher VEI values. Diagrams are after from Walker (1973), Cas and Wright (1988), Newhall and Self (1982). 2. Monogenetic versus polygenetic volcanic systems in a conceptual model after Nemeth and Kereszturi (2015). Letters refer to **a** polygenetic volcano, **b** compound volcano, **c** monogenetic volcano cluster/field, **d** large volume monogenetic volcano with shallow crustal magma storage system, **e** polymagmatic compound monogenetic volcano, **f** polymagmatic simple monogenetic volcano, **g** sensu stricto monogenetic volcano. On the polymagmatic volcano cartoon numbers refer to as (1) deep-fed monogenetic volcano in the ring plain, (2) deep-fed flank volcanoes on the edifice, (3) shallow magma storage-fed small-volume volcano, (4) small volume volcano fed from edifice storage systems. 3. Landscape modifying and landscape forming volcanic systems such as calderas, stratovolcanoes and monogenetic volcanic fields. 4. Magma composition variations as reflection of the petrogenetic processes formed the volcano. 5. Geotectonic systems classified by the recently proposed Earth System approach of De Vries et al. (2018). 6. Volcanic facies architecture determent elements outlined by Nemeth and Palmer (2019). Pyroclastic "Fall" processes tend to generate mantling geometry of their product while pyroclastic "Flow" processes show more valley confined elements. (1a)—deep source chemistry and magma extraction, (1b)—magma source to surface transport, (1c)—magma temporal shallow storage processes, (2)—crystallisation and vesiculation, (3a)—pyroclast transportation through "fall" processes, (3b)—pyroclast transportation through "flow" processes [pyroclastic density currents including pyroclastic flows, ignimbrites, pyroclastic surges], (4a)—deposition processes from "fall", (4b)—deposition from "flow", (5a)—redeposition by laharic (Lh) currents of any type, (5b)—redeposition by volcanic debris avalanches (Vda) of any type. Lava flows are important lithostratigraphic components (Lf)

volcanic facies and facies associations are clearly distinguishable even in ancient settings (Németh and Palmer 2019) and hence can be used to establish the volcanic geoheritage elements.

Volcano types as monogenetic versus polygenetic are already suggestive that some sort of scale of observation has been utilized. Monogenetic volcanoes are about two magnitudes smaller by edifice type, impacted geoenvironment and eruption duration than those considered to be polygenetic (de Silva and Lindsay 2015). While it has not been adequately researched it seems that the boundary between monogenetic and polygenetic volcano types are far more continuous than has previously been considered (Nemeth and Kereszturi 2015). There are very complex volcanoes that still retain monogenetic characteristics from petrogenetic and volcano architecture perspective, but there are also polygenetic volcanoes that are not a lot different from the monogenetic volcanoes. Normally mafic magmas from intraplate (mostly intracontinental) settings produce single, on–off eruptions and generate monogenetic volcanoes, while more evolved magma types tend to form longer lived polygenetic volcanoes like strato- and compound volcanoes. Mafic volcanic rocks are volumetrically more dominant at the surface, hence volcano types fed by mafic magmas are common hence they are in most of the known geoenvironment and geotectonic settings.

Looking at the volcanic architecture and volcanic facies perspective is a scientifically valid approach to identifying volcanic geoheritage elements. Moreover, the volcanic facies approach to define the geoheritage elements of a volcanic terrain can be applied to ancient settings where the original volcanic landforms are heavily modified or already not recognizable.

The volcanic facies approach also can be applied to any volcano type regardless of whether they are dominantly effusive or explosive eruption processes, specific eruption styles (magmatic vs. hydromagmatic/phreatomagmatic), small-volume monogenetic or large volume polymagmatic. This approach could be used as a "*checklist*" of what to look out for when we try to identify the geoheritage elements associated with a volcano. To embrace volcanic facies approach in the volcanic geoheritage element recognition is also advisable as it fits perfectly to any volcano model, independent of its size, composition or geotectonic situation.

Recognition of geoheritage elements of monogenetic versus polygenetic volcanism is a challenge. While individual monogenetic volcanoes are small and the facies restricted to very small spatial scales, larger monogenetic volcanoes commonly form in groups of volcanoes that evolve over millions of years, effected by associated climatic and/or hydrological changes and the overall territory can be a magnitude larger than an average polygenetic volcano. This problem will be a very challenging one when setting

out to establish the geoheritage elements defining the geodiversity or when providing fundamental background data to assist location of geoheritage sites through comparative analysis.

Recognizing the geoheritage elements of the largest landscape forming eruptions can also be difficult. It is common such volcanism produces large volume ($\gg 10$ km^3), macroscopically laterally, very homogeneous ignimbrites associated with mega calderas (10 km + across) (Lindsay et al. 2001; Antonio Naranjo et al. 2018), however, distinct geoheritage elements may be associated with features at microscopic scales (e.g. minerals, xenoliths etc.), or phenomena largely associated with the interaction between the large volume ignimbrites and their geoenvironment (e.g. various peperites, fluid escape pipes, lag breccias etc.). The volcanic geoheritage aspects of such high intensity and landscape forming eruptions also can be linked to extensive tephra fall that can reach well beyond the vicinity of the eruption sources and form a significant geoheritage element in a "*foreign*" geological setting (Breitkreuz et al. 2014).

Volcanic geoheritage elements are very commonly viewed only from modern and active volcanic systems. Volcanic geoheritage of ancient settings are significantly underrated or abandoned (Migon and Pijet-Migon 2016, 2020). In ancient settings erosion commonly removes the majority of the medial and distal sections of a volcanic edifice, over time exhuming the crater and upper conduit facies. In extreme situations, volcanoes reaching a complete eroded phase form unique exposures where the deep interior, or even their magmatic plumbing system, is exposed ready to study or be utilized in geoeducation programs. Volcanic geoheritage of exhumed conduit systems should be considered seriously as they are also part of a volcano/magmatic system and act as a link between the source and the surface manifestation of magmatism.

The interaction of volcanism with the surrounding geoenvironment, or sedimentary basin, create valuable volcanic geoheritage elements (Fig. 9). Volcanic basins, especially along convergent plate margins where arc volcanism takes place, should be taken in account from their volcanic geoheritage perspective. While volcanic edifices erode and gradually diminish after a few millions of years, they may leave behind an exhumed pyroclastic breccia-filled conduit network and associated dyke, sill and other intrusive elements (Lefebvre et al. 2013; Latutrie and Ross 2019). Distal sedimentary basins can accumulate and preserve the volcanic activities well beyond the lifespan of their source volcano forming distinct volcanic-impacted sedimentary basins displaying a typical volcaniclastic sedimentary succession, e.g., the "*Pietra Verde*" in Northern Italy (Budai et al. 2005; Cassinis et al. 2008; Furrer et al. 2008; Németh and Budai 2009; Dunkl et al. 2019). Recognition of such volcanic geoheritage elements of a sedimentary basin is

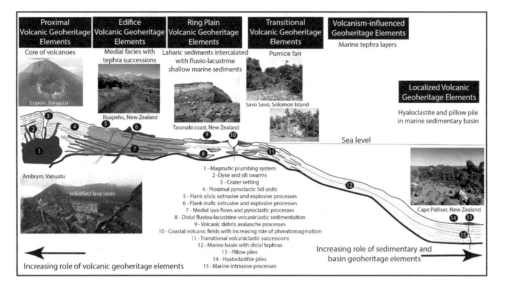

Fig. 9 Interface between volcanoes and the background sedimentary systems/basins in a highly schematic diagram that is not to scale. From the volcanic edifice a decreasing volcanic geoheritage dominance is shown toward the marine basin where "only" volcanism-impacted geoheritage elements can be recognized. Numbers represent key volcanic geology elements associated with the respected zone. Note that the same concept can be applied for eroded volcanic terrains where the core of the volcanic elements is exhumed and exposed. Applying such conceptual framework to identify the key volcanic geoheritage elements can help to link eroded volcanic terrains to modern setting and develop a geologically well designed geoeducation and geotourism program across volcanic geoheritage sites

imperative to recognize and incorporate into a volcanic geoheritage model to ensure it is complete, we call this the holistic vision of volcanism. While we may intuitively think that such scenarios are only or strictly associated with extensive marine basins along various segments of a volcanic arc, the volcanic input of pyroclastic detritus into terrestrial systems is also measurable and plays important part of the terrestrial sedimentary processes (Rees et al. 2018, 2019, 2020). This is particularly valid for intermediate polygenetic volcanoes where the central edifice is commonly surrounded by a complex and very extensive so-called ring plain that forms where sedimentation style is heavily dependent on the intensity of the volcanism, the frequency of the volcanic episodes and the changing climatological parameters affecting the surface water distribution configuration (Zernack et al. 2009, 2011; Németh and Palmer 2019; Zemeny et al. 2021; Zernack 2021; Zernack and Procter 2021).

Volcanism can also form local depressions that function as terrestrial depocenters for sedimentation. Interestingly small depressions, such as maar craters (Lorenz 2007; Christenson et al. 2015), are significant volcanic geoheritage elements (Moufti et al. 2013a, 2015; Yoon 2019; Becerra-Ramirez et al. 2020; Bidias et al. 2020; Megerle 2020b). These small and deep craters can form in terrestrial settings (Graettinger 2018) where they are infilled with terrestrial sediments over tens of thousands of years providing a high-resolution record of past terrestrial environments including climate change and paleoenvironments (Gruber

2007; Nemeth et al. 2008; Zolitschka et al. 2013; Lenz and Wilde 2018; Kovács et al. 2020). In large craters, such as caldera systems, significant volcanic geoheritage elements form as thick (hundreds of metres thick) lacustrine sequences (Branney and Acocella 2015; Christenson et al. 2015; Cattell et al. 2016). These have the potential to feed outbreak floods that can significantly alter the terrestrial environment of a large area such as happened numerous times in the North Island of New Zealand after caldera eruptions expelled large volume of pumiceous material that blocked fluvial networks (Manville 2002; Hodgson and Nairn 2005; Manville et al. 2007; Barker et al. 2021).

Volcanic geoheritage can also be recognized in distal areas far away from volcanoes. These rock successions include broad laharic fans and volcaniclastic fluvio-lacustrine networks associated with outflow drainage networks initiated from central volcanoes (Vallance 2000; Gudmundsson 2015; Procter et al. 2021). For example, in central Colombia major narrow, deep fluvial channels frequently captured lahars from Nevado del Ruiz, Tolima or Cerro Machin volcanoes in the Quaternary (Lowe et al. 1986; Voight 1990; Thouret et al. 2007; Murcia et al. 2008) (Fig. 10). The volcanic geoheritage element of these sudden, large volume sediment inputs into a sedimentary basin is commonly overlooked as volcanic geoheritage.

Intermediate sized eruptions, such as those in many volcanic islands such as in Ambrym, Vanuatu or Savo in Solomon Islands, from volcanic islands can produce alluvial fans that prograde into adjoining the marine basins forming a

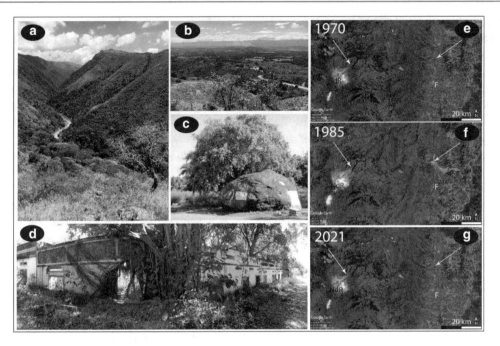

Fig. 10 Laharic geosystems are very important locations commonly associated with "dark" geocultural elements hence their geocultural link can be strong to actual volcanic disasters. Typical laharic facies variation from source to distant area from Nevado del Ruiz volcano and the surrounding catchment areas provide an excellent model how such geosystem can provide the basis of understanding the volcanic geoheritage context of the region. Narrow fluvial arteries (**a**) feed volcaniclastic sediments into the local terrestrial basin of the Magdalena River (**b**). In 1985 November a dramatic lahar event carried volcanic debris over 60 km length to the alluvial basin, capable to successfully move mega-blocks over 10 m in diameter to such distance (**c**) and inundate Almero township killing over 22,000 people. The laharic fan development is clearly visible on the pre-lahar January 1970 (**e**), the immediately after-lahar December 1985 (**f**) and the current July 2021 GoogleEarth Pro satellite images. Yellow arrows point to the location of Armero township. White arrows point to the initiating point of the valley channelized the lahars in 1985 from Nevado del Ruiz (NdR). Note the other Quaternary volcaniclastic fans in the region (**f**) providing evidences of the globally significant scale of lahar processes and their depositional impact on the terrestrial environment. In conjunction with the "dark" geocultural elements as a significant volcanic disaster, the region is the perfect location to look at it as a "best" and "most" in lahar-associated volcanic geology

unique geoenvironment (Petterson et al. 2003; Németh et al. 2009). The volcaniclastic volume contributes to landmass growth and probably plays an important role in the landscape evolution especially where volcanism is frequent. Active volcanic arc regions, such as the Taupo Volcanic Zone in the North Island of New Zealand, is an example of where volcanism has impacted landscape evolution (Manville 2002). Here active volcanism has contributed primary volcanic and volcaniclastic sediment into the surrounding landscape for c.1.8My (Alloway et al. 2005; Pillans et al. 2005). In that time long sedimentary transport arteries have evolved facilitating the transport of pumiceous deposits to broad coastal plains and beyond. This frequent volcanic activity and the associated large volume pyroclast-producing volcanic events have played a significant role in the development of the resultant geoenvironments. Hence, they should be considered an important volcanic geoheritage element.

Volcanism commonly produces large volume of lava on the surface. Lava flows and their accompanying surface textures are distinctive and recognizable. They are directly related to the processes that generated them and the physico-chemical conditions that existed at the source of the magma (Kilburn 2000). Recognizing the volcanic geoheritage of lava flow fields should lead to a search for specific facies of the flow field that relate to the source. In extreme cases, extensive lava flow fields form large igneous provinces, such as those in the Karoo, Columbia River Basalt or Deccan (Bryan et al. 2010; de Silva and Lindsay 2015; Sheth 2018). These all, cover thousands of km^2 and act as landscape-forming volcanic geoheritage elements. The best approach to understanding the volcanic geoheritage of these mega-features is to develop a "portfolio" based upon the observed volcanic facies.

Overall, we conclude the volcanic geoheritage of Earth is an absolute element and should be independent of the purpose, goal, scope or geocultural perspective from which it looked at. Volcanic geoheritage elements need to focus on the best possible volcano model and to recognize the features associated with it in a specific region. Identified sites can then be used to evaluate their relative significance through a preselected purpose-dependent scale and scope.

4 Identification and Comparison of Volcanic Geoheritage Sites and Developing a Volcanic Geodiversity Estimate

A general workflow outlined in several studies recommends the steps to follow to develop a volcanic geodiversity estimate for a region (Brocx and Semeniuk 2007, 2019; Brocx et al. 2021). The model proposed here consists of at least three major stages. In stage one it is necessary to define the purpose of the assessment that can be linked to the recognition of the various landscape elements of the volcanic geoheritage identified, its special variations and their significance to a pre-defined scope and scale. After setting up the method the selection of geosites and/or geomorphosites should follow. This work should be conducted after the recognition of volcanic geoheritage elements in region have been studied and following the conceptual framework outlined in the previous sections. After completion of this stage, it is possible to locate key volcanic geosites. These key sites are the geoheritage sites, that can also be named as geosites or geomorphosites depending on whether the main emphasis of the valorization is geology or geomorphology. Then the identification of various volcanic geoheritage elements suitable to locate volcanic geosites can be undertaken. This will involve identifying those places that contribute the most to our understanding of the volcanism in the study area, to see the best examples of the identified eruption styles, volcano types, volcano-geoenvironment interaction places or impacts to the surrounding biotic and abiotic nature including the human society. In the final stage of this progression attention should be paid to the management and conservation policies that need to be developed to preserve the key geoheritage features that form the basis for any geotourism and geoeducation initiatives that follow. It is imperative the most significant geoheritage elements should be identified clearly and linked to the original purpose of such research (Fig. 11).

Geosite valorization, in general, is a process where the "*most and best*" of the identified geoheritage elements are selected. The main aim is to develop a workflow model that produces objective and reproduceable results. In many cases subjectivity can be difficult to abandon so it is considered an achievement if the subjectivity can be reduced to a level where the resulting selection and associated valorization results in more or less the same results for each new study on the same location. Recently a series of works has been published proposing some sort of geoheritage toolkit that helps the user to work out the strategy and the actual valorization of geosites (Brocx and Semeniuk 2007; Brocx et al. 2021). The toolkit was tailored specifically to volcanoes to systematically identify and assess sites of geoheritage significance (Brocx et al. 2021). This toolkit is based on the

identification of the (1) conceptual categories of sites of volcanic geoheritage significance, (2) the scale of volcanic geoheritage features recognized and (3) the recognition of the volcano (or volcanic geoheritage features) significance. This method provides a very good simple workflow to follow but also generate some ambiguity or imprecise categorization. For this reason, it is suggested the toolkit needs significant revision that introduces a more precise fit to the most recent scientifically backed volcano models (de Silva and Lindsay 2015; Martí et al. 2018; Németh and Palmer 2019). The first two steps of the toolkit involve the identification of the volcanic geoheritage elements. In this aspect, it is suggested the geotectonic concept is incorporated in more detailed way, similar to that suggested for the Earth System geoheritage recognition (De Vries et al. 2018). In the main part of the geoheritage element recognition of the magma to source perspective, petrogenetic aspects, volcano model application (e.g., monogenetic vs. polygenetic volcano types), the volcanic facies model (e.g., for both volcano types but also for the lava flow fields) and the interface recognition where the volcanism interacted with the geoenvironment (e.g., volcaniclastic sedimentation etc.). As outlined previously, this is a very important stage as it will form the basis of any valorization and site selection. In the third step each of the recognized volcanic geoheritage elements should be measured against the conceptual category suggested also by Brocx and Semeniuk (2007). Their categories focus either the product of volcanism such as the geoform of modern volcanoes, the preserved products of ancient settings and active volcanic sites. In the fourth step the scale of the geoheritage features should be determined and valorized (e.g., using its representativeness) while in the fifth step the significance of the volcanic geoheritage features should be evaluated. The final outcome of the entire valorization process should be a decision on the level of conservation or management that is applied to the location.

While this toolkit sounds like a reasonable first order proxy, the details contain some issues in particular the definition of the significance of the geoheritage feature. The techniques most commonly used are almost exclusively linked to an artificially defined spatial value which may be local, regional or global (Brocx and Semeniuk 2007). Here it is suggested that the significance of specific feature should initially be referenced to the scale of the identified volcanic geoheritage feature. For instance, the significance of a volcanic geoheritage feature identified in relationship with a monogenetic volcano should be measured against the individual feature itself, i.e. within a single scoria cone or tuff ring (e.g., this could translate to local), across the volcanic field (e.g., this could be regional scale) or volcanic field to volcanic field within a geotectonic situation (e.g., this could translate to international scale) or across the entire globe's all

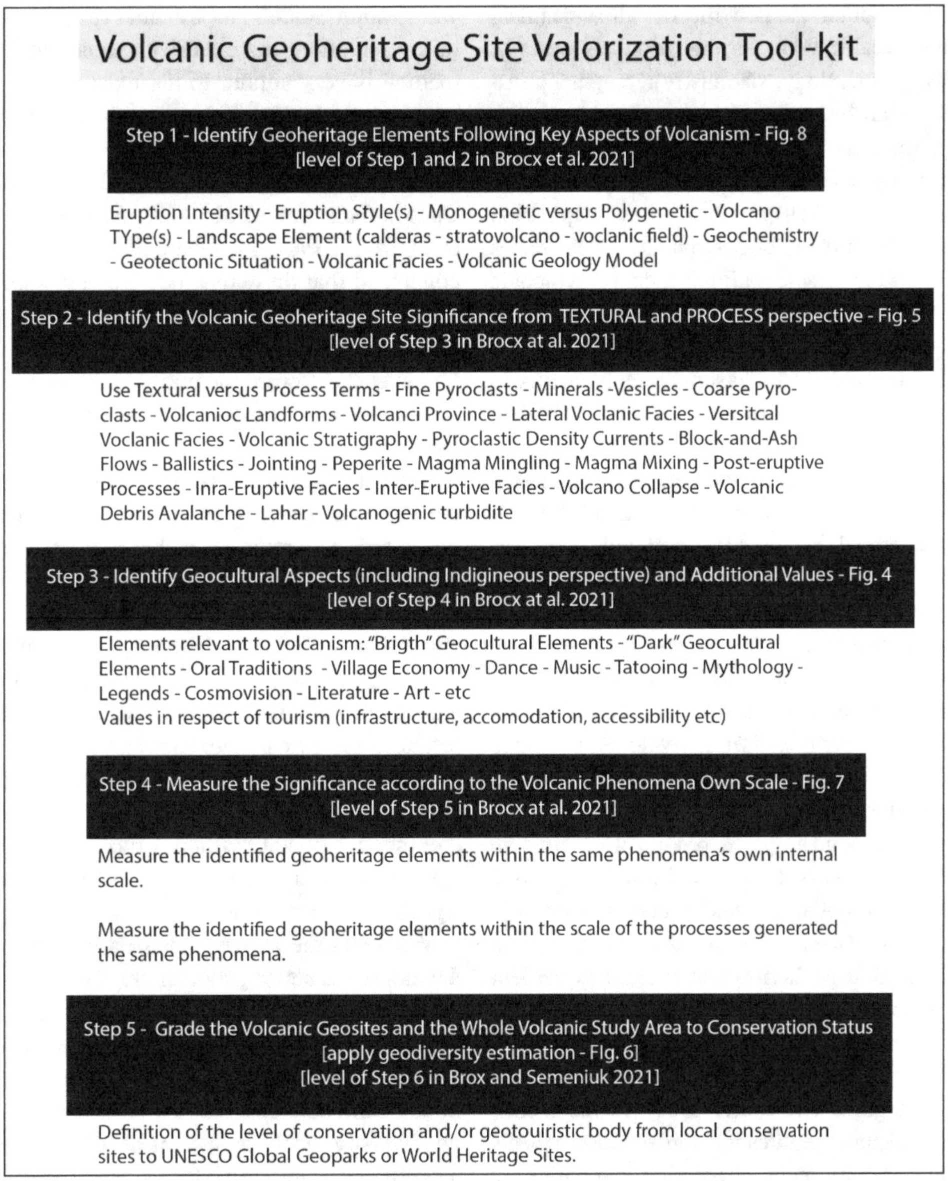

Volcanic Geoheritage Site Valorization Tool-kit

Step 1 - Identify Geoheritage Elements Following Key Aspects of Volcanism - Fig. 8
[level of Step 1 and 2 in Brocx et al. 2021]

Eruption Intensity - Eruption Style(s) - Monogenetic versus Polygenetic - Volcano TYpe(s) - Landscape Element (calderas - stratovolcano - voclanic field) - Geochemistry - Geotectonic Situation - Volcanic Facies - Volcanic Geology Model

Step 2 - Identify the Volcanic Geoheritage Site Significance from TEXTURAL and PROCESS perspective - Fig. 5
[level of Step 3 in Brocx at al. 2021]

Use Textural versus Process Terms - Fine Pyroclasts - Minerals -Vesicles - Coarse Pyroclasts - Volcanioc Landforms - Volcanci Province - Lateral Voclanic Facies - Versitcal Voclanic Facies - Volcanic Stratigraphy - Pyroclastic Density Currents - Block-and-Ash Flows - Ballistics - Jointing - Peperite - Magma Mingling - Magma Mixing - Post-eruptive Processes - Inra-Eruptive Facies - Inter-Eruptive Facies - Volcano Collapse - Volcanic Debris Avalanche - Lahar - Volcanogenic turbidite

Step 3 - Identify Geocultural Aspects (including Indigineous perspective) and Additional Values - Fig. 4
[level of Step 4 in Brocx at al. 2021]

Elements relevant to volcanism:"Brigth" Geocultural Elements - "Dark" Geocultural Elements - Oral Traditions - Village Economy - Dance - Music - Tatooing - Mythology - Legends - Cosmovision - Literature - Art - etc
Values in respect of tourism (infrastructure, accomodation, accessibility etc)

Step 4 - Measure the Significance according to the Volcanic Phenomena Own Scale - Fig. 7
[level of Step 5 in Brocx at al. 2021]

Measure the identified geoheritage elements within the same phenomena's own internal scale.

Measure the identified geoheritage elements within the scale of the processes generated the same phenomena.

Step 5 - Grade the Volcanic Geosites and the Whole Volcanic Study Area to Conservation Status
[apply geodiversity estimation - Flg. 6]
[level of Step 6 in Brox and Semeniuk 2021]

Definition of the level of conservation and/or geotouiristic body from local conservation sites to UNESCO Global Geoparks or World Heritage Sites.

Fig. 11 Modified volcanic geoheritage toolkit that is based on the concept of Brocx et al (2021). One of the main differences in this modified version that it takes the scale measured to the specific volcanic geology problems, process and measurable features. It is also better fit to the volcanic facies models and the interaction between volcanism and the background sedimentary processes. The following "Steps" suggested to follow: 1. Identification of volcanic geoheritage elements —some key elements listed. 2. Identification of the significance of volcanic geoheritage elements including textural and process associated features volcanism—some key approaches listed. 3. Identification of the geocultural elements (including indigenous aspects) associated with the identified volcanic geoheritage elements. Recognition of key additional values from the perspective of the purpose of the valorization —some key element listed. 4. Measure the significance of the volcanic geoheritage elements within the volcanic phenomena common scales. 5. Grade the identified volcanic geosites and create a systematics for the general purpose of the analysis. Define the level of conservation and protection. Please note that his is a theoretical approach and the basic concepts can be fine-tuned to the volcanic terrain under investigation

volcanic fields (e.g., this could be the global scale equivalent category). In a similar way we could apply the same volcano geology-based logic to polygenetic volcanoes such as (1) within the same volcano, (2) within the same volcanic province and/or volcanotectonic regime and (3) across the globe.

The scale of the geoheritage feature (step 4) is something that is difficult to comprehend, but intuitively a category that is likely related to the processes forming the various identified geoheritage elements and their scale where the appropriate evidence could be identified. This step could be further enhanced using a volcano model that is linked more directly to the conceptual framework of all volcanism manifest at various scales with various products such as magma generation (mineralogy, chemistry), magma transportation (various microtextural features, magma vesiculation and fragmentation (bubble textures, microlite distribution patters, pyroclast shapes and vesicularity etc., pyroclasts morphology), transportation and deposition (bedding features, transport indicators, outcrop-scale facies association, field-wide associations) and volcanic edifice/complex/geoform evolution (facies associations, 3D architecture, landscape scale features). It is evident that the suggested line of observable and measurable features link to specific observation scales that are more or less aligned to the suggested macro, mezzo and micro scale approach by Brocx and Semeniuk (2007). The advantage of the approach suggested here is that this method directly links to the volcanic processes generated by the identified volcanic geoheritage elements.

A recent study of the Garrotxa Volcanic Field, Catalunya, provided a very workable method to test how such an approach may work (Planaguma and Marti 2020). The presented method is not structured exactly as suggested in this work, but in the geosite identification and valorization stage naturally follow similar techniques outlined in this chapter. The volcanic geoheritage recognition has been based on the recognition of volcanic deposit types, produced by effusive and explosive processes associated with Strombolian-style explosive, violent Strombolian explosive and phreatomagmatic explosive eruptions (Planaguma and Marti 2020). From the list of volcanic geoheritage features it is evident that the studied volcanic field hosts most of the expected geoforms that form by the eruptive processes identified by volcanological research targeting volcanic fields over the past 100 years (Planaguma and Marti 2020). In the identification of the key volcanic geosites the conservation aspects played an important role following the notion that without a good initial conservation plan all the geosite inventory builder or volcanic geoheritage element documentation would just remain a theoretical work without significant effect on the planning and development of the region. From this, mostly economy-driven reason the volcanic geosite identification locates the most interesting outcrops and illustrate the great variety of the eruptive products in the field. The selected volcanic geosites need to, therefore, represent the main and most significant elements of the volcanic field. Additional geosites were selected mostly from their "additional value" perspective for activities, conditions, accessibility, land-use status, immediate surroundings, space and fragility (Planaguma and Marti 2020). The many aspects were exclusively centered around conservation measuring the current conservations state, the site abundance or uniqueness, its type, its link to other natural phenomena and diversity elements. Each track followed a three-point

valorization model, and the final result represented the level of conservation interest (Planaguma and Marti 2020). This method is very similar to the most common geosite assessment methods (GAMs) used elsewhere in other regions, predominantly from a geotouristism point of view (Vujicic et al. 2011; Moufti et al. 2013b; Bratic et al. 2020; Cuevas-Gonzalez et al. 2020; Szepesi et al. 2020; Ibanez et al. 2021; Pal and Albert 2021). In summary it can be concluded that these methods are very specifically designed for a specific purpose, namely geotouristism and less commonly geoconservation purposes. Either way the valorization is biased toward the utility values of the sites and tend to be detached from the geoheritage site volcanic geoheritage values.

Here it is suggested a modified toolkit be used to valorize the volcanic geoheritage sites (Fig. 11). This toolkit would put a greater emphasis on the correct identification of the geotectonic situation, Earth System position, volcano type recognition as well as the application of the volcano model and volcanic facies to define key elements of the processes resulting in specific volcanic geoforms. In short, the higher and more precise usage of volcanic science applied to establish the volcanic geoheritage elements is recommended to generate a more science-aligned volcanic geoheritage model to identify key volcanic geosites. The updated toolkit then should operate within a more realistic conceptual categories, a better internal scale and volcanic process-defined significance categories.

Applying the above principals, we also can get closer to developing a better geodiversity (Gray 2018a, b; Zwoliński et al. 2018; Fox et al. 2020; Dias et al. 2021; Wolniewicz 2021) recognition method applicable to volcanic features (Dóniz-Páez et al. 2020; Quesada-Román and Pérez-Umaña 2020; Guilbaud et al. 2021; Vörös et al. 2021). Defining geodiversity itself is a subject that is under debate and recently evolving fast (Brocx and Semeniuk 2019, 2020; Gray and Gordon 2020), hence its application to volcanic regions still in infantry. For geodiversity estimates for volcanic regions the best possible available volcanology map is needed. A volcanology map should fulfil the expectations of geological map production with an additional feature that it is also specific to what a volcanic system can produce. Volcanic processes occur, in general, much faster than normal sedimentary processes, hence a volcanic terrain will contain a larger number of geological features that may change over much shorter distances at greater rates than other geological features such as those in a siliciclastic marine sedimentary system (Németh and Palmer 2019). Hence, the volcanic geodiversity is expected to be large in a given area in comparison to other normal sedimentary successions. This may not be visible on a standard geological map at a scale 1–50,000 or smaller. The problem we face here is similar to the problem of geological mapping and to find the best scale to visualize, in map format, volcanic eruptive episodes. This paradox can be resolved by using volcanic facies to identify volcanic geoheritage elements. The scale of the study sites in a typical polygenetic volcano such a strato, compound or caldera volcano allows the typical volcanic facies and volcano geology concept to adapt for geodiversity estimates. Within volcanic fields the size of individual volcanoes could be too small to capture appropriately the various geodiversity elements associated with monogenetic volcanism. On other hand, however, geodiversity estimates can be examined in a larger scale that fits the spatial scale of a typical monogenetic volcano (Smith and Németh 2017). In this respect the scope of generating geodiversity estimates for monogenetic volcanism should be identified and measured to the usual spatial scales of such volcanoes. This scale problem, however, will likely affect the comparison or fitting into a single geodiversity estimate map a volcanic terrain that consists of both monogenetic and polygenetic volcano types. To test this problem and develop a simple and workable method to handle this spatial discrepancy has not been done yet and signifies a knowledge gap that future research should target. In other hand, small monogenetic volcanoes can be treated as a single geodiversity "source" based on detailed studies of identification of the number and weight of geoheritage elements associated with the specific volcano types identified.

5 Link Between Volcanic Geoheritage and Other Geoheritage Elements

Following the conceptual framework of volcanic geoheritage outlined in previous sections, it can be expected that there will be a connection of the identified volcanic geoheritage elements and the non-volcanic geoheritage elements associated with it. Volcanoes are part of complex sedimentary systems, geoenvironments and geotectonic settings (Németh and Palmer 2019), hence they are a vital part of the overall geoheritage of any region. Separating volcanic geoheritage from other heritage elements is sensible only if (1) the studied terrain is dominated by volcanic geoforms and volcanic eruptive products or (2) the study is specifically targeted to understand the volcanic geoheritage elements and utilize them for other purposes such as conservation, geotourism or geoeducation. The separation of volcanic geoheritage and their treatment as a separate entity could lead to similar data and map sets common in the so-called thematic map series dealing with specific geospatial problems. The volcanic geoheritage, in this perspective, should be treated as a vital part of the total geoheritage scene and provide clear interlinkages to the other geoheritage elements it is embedded in. It is very important to follow the volcano geology framework outlined in previous sections as volcanoes

commonly produce large volumes of eruptive products that eventually accumulate in sedimentary basins and provide unique scenarios such as various siliciclastic sedimentary rocks formed in sedimentary basins influenced and impacted by various types of volcanism. The role of volcanism on such sedimentary basin evolution is likely associated with those systems where prolonged periods of volcanism produces a relatively steady volume of volcanic detritus into a sedimentary basin (Németh and Palmer 2019). In such scenarios the geoheritage elements will be associated by normal sedimentary processes but the appearance of the resulting rocks could be distinctly different due to the volcanic origin of their constituent elements. Strictly speaking in such context, the identified geoheritage elements is not a volcanic one. Many greywacke basins would fall into such a category, where the sedimentary basin sedimentation is influenced by volcanism and a specific type of greywacke composed of volcanic lithics in a sand or finer grain sizes is found (Challis 1960; Roser and Grapes 1990; Laumonier 1998; Benedek et al. 2001; Floyd 2001; Bennouna et al. 2004; Bandopadhyay 2005) (Fig. 9). The geoheritage elements of such a scenario are more likely associated with the deep marine sedimentary processes than the distal volcanism itself. There will be a transition from deposits dominated by sporadic seafloor volcanism to medial eruptive products intercalated within siliciclastic sedimentary successions. In this perspective such locations can and should be viewed as a volcanic geoheritage element (Fig. 9). Moreover, a normal greywacke basin produces a thick pile of very monotonous rocks often used for geological terrain recognition (Michaux et al. 2018) that can change their appearance and macro and microtextures when such volcanic interbeds are present, hence such sites can form significant landscape features that may stand out and can be utilized for geotouristism or geoeducation purposes and even be part of a focused conservation effort. To define the boundary between when we call a setting a volcanic geoheritage element or a normal geoheritage element should follow the processes associated with volcaniclastic sedimentation such as the recognition and the number of incidents of primary eruption-fed pyroclastic successions within the non-volcanic background sediment. The volcanic facies model recognizes geoheritage elements of mixed volcanic and normal sedimentary settings are important and should be treated with great care as they can be utilized for geoeducation and geotouristism.

6 Geocultural Aspects on Volcanic Geoheritage

It is a common and recurrent argument is that the geoheritage elements should contain the geocultural aspects of the recognized features (Reynard and Giusti 2018; Kubalikova

2020). In the conceptual framework presented here it is suggested that the geoheritage elements be distinctly separated from any geocultural aspects including indigenous or alternative cosmovisions. The geocultural and the indigenous aspects of geoheritage sites, however, should be acknowledged in the geosite recognition and/or the valorization of such sites (Gravis et al. 2017, 2020). In addition, indigenous values are recently has been considered as measurable aspects of geoheritage, however the way how those should be included in any valorization method is currently not known. Perhaps indigenous values could play vital roles in geoconservation where living indigenous cultures act on the land or where archaeological sites are abundant (Turner 2013; Lim 2014; Clifford and Semeniuk 2019; Lewis 2020). This process, however, is more connected with the establishment of a complex valorization structure for geotouristism or geoeducation and should also incorporate geoconservation strategies. Currently geosite assessment methods underutilize the indigenous aspects of geosites such as exiting culturally significant sites or oral traditions associated with a region (Fepuleai et al. 2017, 2021; Reynard and Giusti 2018). Geodiversity estimates where distinct geoheritage elements are evaluated and counted within their spatial extent they commonly yield average or below values across a region especially when geological features are evenly distributed across the known geological assets. In such regions geosites that are significant from a geotouristism or geoeducation perspective should be incorporated into the available geocultural dataset, including the region's indigenous human settlement history. Within the framework of volcanic geoheritage, rich geocultural aspects of specific sites can be recognized either by the positive affect volcanism had on the human societal evolution or the negative, often destructive power, volcanism posed on the human beings through volcanic disasters (Cronin and Neall 2000; Scarlett and Riede 2019). The dark geocultural impact is a measurable fact that can be collated from participatory methods applied to understanding the oral traditions linked to volcanism as well as being part of cultural activities of everyday life and ritual-driven activities (Nunn et al. 2006, 2019; Cashman and Cronin 2008; Cashman and Giordano 2008; De Benedetti et al. 2008; Swanson 2008; Németh and Cronin 2009; Donovan 2010; Scarlett and Riede 2019; Wilkie et al. 2020). Impact of volcanism on the society manifest very diverse cultural responses hence volcanism commonly need to look from the "*living with volcanoes*" aspect that all together can form an intact and internally coherent knowledge system, cultural traditions or living practices all together can form a distinct geocultural aspect of volcanism (Kelman and Mather 2008). To explore and harvest the accumulated knowledge for a purpose of developing strategies to preserve this geocultural entity interconnected geoscientists are needed whom able to

find the link among various knowledge systems and able to be part of participatory methods and co-development of development toward geoeducation, geoconservation or geotourism goals (Cronin et al. 2004a, b; Nahuelhual et al. 2016; Petterson 2019; Marin et al. 2020; Fepuleai et al. 2021). It is recommended a more structured approach and standalone treatment of the geocultural aspects of volcanism be developed to enable better comparison across regions and human societies. It is also a matter for debate how we treat indigenous knowledge and record natural phenomena associated with volcanism. One can argue that indigenous knowledge extraction differs from a so-called western data collection and could contain knowledge elements that differ from common western scientific knowledge. Such an issue is a real problem in regions where the western scientific knowledge of a volcanic terrain is limited, for instance by the pure lack of scientific research on the features. In such places it is particularly important to incorporate the traditional knowledge about volcanism and to utilize it within the earlier outlined volcanic geoheritage framework. Ideally such an approach should be followed in every volcanic terrain with multicultural and indigenous links.

In summary, geocultural aspects are additional values that can be decisive in geotouristism, geoconservation, rural and urban planning and geoeducation (Dóniz-Páez et al. 2011; Paulo et al. 2014; Riguccio et al. 2015; Zangmo et al. 2017; Megerssa et al. 2019; Beltran-Yanes et al. 2020; Hlusek 2020; Schwartz-Marin et al. 2020; Vizuete et al. 2020; Yepez Noboa 2020). Geocultural values play an important role in finding a sustainable approach for human society to live with volcanoes. As volcanoes equally provide a lifeline to human beings (e.g., good soils, agriculture, spiritual aspects) as well as destruction (e.g., volcanic catastrophes) the interaction between human society and volcanism function as a key element in understanding our environment hence it has huge heritage value. Especially as frequently active volcanoes such as polygenetic stratovolcanoes on convergent plate margin settings function create landforms the human population had to learn to live with. In this symbiotic relationship volcanism can be deeply embedded not only in the cultural practices and legends but also in everyday life activities. To a certain extent, human migrations are triggered, or evolution of civilizations heavily altered by volcanic eruptions hence volcanism plays a significant role in the development of a cultural landscape over a volcanic terrain (Plunket and Urunuela 2005; Pardo et al. 2015, 2021). It is probably a logical conclusion that geocultural aspect of volcanism in this regard are governed by the natural processes of volcanism that determine how societal evolution takes place. Determining the effect of the direct and unseparatable elements of volcanism on a region is a difficult problem. Landforms and geological processes have often shaped the cultural evolution of entire regions hence one can argue that this symbiotic interrelationship between volcanism and society should be included in the volcanic geoheritage element identification (Balmuth et al. 2005; Cecioni and Pineda 2006; Streeter et al. 2012; Black et al. 2015; Zeidler 2016; Oppenheimer et al. 2018). Here it is suggested that we separate this element as volcanism could have taken its course governed by the natural processes regardless of the existence of any society nearby and in sensu stricto volcanic geoheritage is the heritage of the natural processes generated them rather than a reflection of the human perception, cultural activity, socio-economic development. For this reason, the usage to treat this interface between natural (abiotic) and human (societal) heritage elements separately and define it as geocultural element is a very practical notion. In this way we can separate the individual elements associated with volcanism from its societal perspective and impact to make a clearer and easier to develop valorization suitable for geoheritage site identification. The validity of this is shown very well with the intraplate monogenetic Auckland volcanic field. This volcanic field consists of at least 53 individual monogenetic volcanoes with the majority initiated by a brief explosive magma-water interaction phase that changed to more magmatic explosive and effusive stages later in their eruption (Kereszturi et al. 2014, 2017; Hopkins et al. 2021). This trend is clearly evident in those regions where the available external water diminished quickly during the eruption producing a higher magma volume and rate resulting in volcanic landforms consisting of large complex scoria cones with sizeable craters. These scoria cone complexes are commonly located in slightly elevated regions and form visible landforms about 100 m above the coastal plains and the nearby harbors (Kereszturi and Nemeth 2016). Such scenes would have captured the attention of early Maori settlers who utilized them as defendable natural fortresses (Davidson 1993). The surrounding ash plains provided excellent volcanic soils for early horticulture and agriculture supporting Maori communities and their early urbanization for about 300 years after their arrival in Aotearoa (Davidson 2011). While estimates on total population associated with fortified cones provide large numbers of over several thousands, some archeology-based estimates suggests nearly a magnitude less, around several hundreds of population within association to a single cone (Fox 1983). Today, these scoria cone landforms are iconic landmarks, and they are strong geocultural sites linking Maori cultural and societal practices to their land. These scoria cones, while they are visually attractive, rarely contain exposed outcrops to see their geological buildup, but even if we have such outcrops, the scoria cones itself are just like any other scoria cones anywhere and in any geotectonic settings. The volcanic geoheritage elements of such scoria cones are restricted to basic geological features and not particularly unique or

outstanding (Nemeth et al. 2021b). However, the indigenous and geocultural aspects of the scoria cones provide significant values and give extra protection status from a conservation perspective and in recent time from geotourism aspects. These cones are now under special conservation and land use policies, an excellent outcome of the heritage notion [https://www.maunga.nz/]. However, in the southern part of greater Auckland city, the geological conditions allowed for the formation of monogenetic volcanoes that are far more interesting and unique from a volcanological perspective (Agustin-Flores et al. 2014), in fact these are examples of the type of explosive eruptions that could occur to in now highly populated area of Auckland in the near future (Nemeth et al. 2021b). The volcanic architecture, their volcanic rocks and the information we could gain from these volcanoes are fundamental to our better understanding of monogenetic volcanism in Auckland and beyond. The volcanic geoheritage elements are enhanced by their geocultural aspects making these locations geologically more valuable because their geodiversity is greater than other volcanoes currently argued as high geoheritage value sites. Currently it is very difficult to argue they come under better protection policies. This is further exacerbated by strong urban growth pressure which will see valuable geocultural sites vanishing with rapid speed and without a policy change it is unlikely that they can be rescued for the future (Nemeth et al. 2021b). While significant effort over recent years has demonstrated the volcanic geoheritage elements of these sites and how those could be built into volcanic geosite identification methods, in practical sense this work has fallen on deaf ears and had little impact on policymaking.

The Auckland case also highlights the problem that volcanic geoheritage elements are underrated despite a better volcanic hazard model being successfully created and communicated at the UNESCO IGCP 679 project titled "Geoheritage for Geohazard Resilience". The gradually and rapidly vanishing sites in Auckland are especially important locations for such programs to adopt.

7 Geoconservation in the Light of Global Perspective of Volcanic Geoheritage

Above we outlined the importance of the geocultural elements of volcanism and demonstrated that clear separation of volcanic geoheritage elements and their geocultural aspects is needed to better identify their fundamental principles. This notion is probably a valid conceptual framework for geoconservation as well. From a volcanic geoheritage element perspective, geoconservation should clearly target the identified geoheritage elements. Geoconservation itself should serve as an internally coherent method to identify the volcanic geoheritage sites that are from the pure volcanic geoheritage perspective considered to be unique. To do this, the scope and scale concept is seemingly a valid approach. The scope in this respect should be something that embraces the identified volcanic geoheritage elements, for instance base surge beds in a tuff ring. The scale in this aspect is defined by (1) the hosting volcanic feature (like a tuff ring in the example before) hosting the volcanic geoheritage element, (2) the volcanic field itself where the tuff ring is located, (3) the volcanic fields that are part of the same volcanic province or geotectonic setting and (4) the global scene of Earth. As space exploration is bringing more and more new data about planetary geology, this logic could later on be expanded to an interplanetary scale.

The logical architecture of conservation will create a workable and transparent, scientifically based systematics of volcanic geoheritage that can provide good raw data for valorization, applying all those methods so far developed in many places (e.g., Geosite Assessment Methods- GAMs methods). The valorization then could also apply advanced technologies such as drone, LiDAR or other remote sensing data for data acquisition or GIS technologies for geospatial analysis of identified values to create thematic maps such as geoheritage intensity, geoconservation susceptibility or geoeducation value.

8 Volcanic Geoheritage as Basis of Geopark Concept

Volcanic geoheritage is a significant and unique element of potential conservation strategies as outlined in previous sections (Nemeth et al. 2017). Volcanic geoheritage can experienced by the fascinating processes of volcanic eruptions, the dark geocultural aspects through past and current volcanic disasters, as well as the huge impact volcanoes have had on human society. Together these make volcanoes an important element to be included in any formal conservation strategies and broader geoeducation programs. In addition, volcanoes and their volcanic geoheritage provides a foundation on which to utilize those geoheritage elements in formal and informal geoeducation programs targeting the aim of making society more resilient for volcanic disasters and better understanding of the role of volcanism in landscape evolution (Migon and Pijet-Migon 2016; Rapprich et al. 2017; Szepesi et al. 2017; Fepuleai and Nemeth 2019; Dóniz-Páez et al. 2020). Geoparks are a relatively new concept that became a globally significant avenue to promote Earth Sciences to a broader community, to act as engine of geotourism and something that can contribute to the local development of a region, hence bear values in respect of economical sustainability (Henriques et al. 2011; Lim 2014; Ruban 2017; Escorihuela 2018; Macadam 2018). The UNESCO World Heritage network recently outlined

and identified the knowledge gaps within the framework of the World Heritage site listing while also making significant effort to promote geosciences through a global network of geoparks [https://en.unesco.org/global-geoparks]. While the World Heritage concept is based on the "*most and best*" or outstanding universal values concept, geoparks are more like a "*bottom to top*" approach to promote geosciences and protect their geoheritage elements (Henriques et al. 2011; Turner 2013; Henriques and Brilha 2017; Catana and Brilha 2020). Geoparks are normally expected to grow out from local community works and form part of a sustainable development design where the abiotic nature forms a core of such works (Brilha 2018b; Catana and Brilha 2020). In this concept ecosystem services can play an important role to define, pronounce and promote the "*services*" abiotic nature provides for the human society (Gray 2012; Gordon and Barron 2013; Gray 2018a, c; Fox et al. 2020). This also can be expressed in the form of direct economic figures and feed concepts to rural and urban development planning. Geoparks in this framework can form a scientifically well-designed and supported avenue where geoheritage elements form the base of conservation and education, largely serving the goal of transfer knowledge to everyone about our abiotic nature.

The level of recognition, which is somehow associated with the scope, scale and significance of the identified geoheritage elements can be expressed in the formal hierarchy of geoparks from locally protected conservation lands to be part of the UNESCO Global Geopark Network [https://en.unesco.org/global-geoparks]. Among UNESCO Global Geoparks many of the properties are strongly linked to some single or set of volcanic geoheritage elements or some additional component which would fall in the geocultural aspects of the region. Recently UNESCO Global Geoparks with pure volcanic geoheritage as a center of their core protection and education program are also common.

Geoparks with strong link to volcano geology are the perfect avenues to disseminate geological concepts associated with volcanic geohazards. The variety of geoparks in the global scale also provide an opportunity to interlink geoparks with similar volcanic geological geoheritage elements. Such method has been proposed in a far more direct way such as the European Volcano Road (Abratis et al. 2015) or the Pannonian Volcano Road (Harangi 2014) as potential examples. Volcanic geoheritage of extinct volcanoes is a common basis of geotourism development and subject of geoconservationa cross continental Europe (Migon and Pijet-Migon 2016; Pijet-Migon 2016; Pijet-Migon and Migon 2019, 2020; Megerle 2020a). The benefit is to focus on educationally well-designed interlinkages so that communities living in currently inactive volcanic regions can get direct knowledge from those similar volcanic geoheritage regions active today. This can help the people to embrace and understand the volcanic processes as well as the landscape evolution perspective of volcanoes. This is particularly important when we are looking at volcanic geoheritage properties of old and young settings. For instance, in regions like Auckland the young age of the volcanoes provide little opportunity for the people to "look inside" their volcanoes, hence transferring knowledge such as magma fragmentation or conduit processes are problematic. While interlinking a place like Auckland with locations where eruptive products of similar but much older volcanics, such as those in Central Europe exists, can help to understand what to envision, and more importantly, what to expect from a future eruption where magma fragmentation might occur. There is a huge age range, compositional, geodynamic and geoenvironmental settings within monogenetic volcanic fields formed across the Earth in the last 600 million years (Nemeth 2016a, b, 2017). These provide opportunity to study similar volcanic systems, with different exposure levels, focusing on different aspects of the same style of volcanism. Similar interlinkages are available for polygenetic volcanoes such as stratovolcanoes, or caldera volcanoes. To date, there has been very little direct attempts to do this.

The recognition of a region as geopark needs strong community support, strong scientific background and knowledge, and very clear valorization methods to see where the real values and what are the key geosites. In this work, the valorization from tourism or conservation perspective can play important role in manifesting the geoheritage elements in a workable framework. Geocultural aspects play key roles, especially in indigenous territories, where oral traditions, legends, cultural activities also exist and their preservation as well as passage through generations could function as a driving force to generate a geopark. Geoparks in this aspect should reflect true transdisciplinary nature and explore the identified geoheritage elements link, association or influence on the archaeological aspects, traditions, and contemporary cultural activities, including lifestyles or living practices (e.g., village culture, agriculture, culinary traditions etc.).

Overall, geoparks could be the engines of sustainable development and a contributor to inclusive conservation and education methods that point well beyond of the geological heritage itself. Volcanic geoheritage through, the experiences of volcanism in the past, is also a significant part of the conceptual architecture of geoparks.

9 Conclusion

In summary, volcanic geoheritage elements are those directly linked to the physical and chemical processes responsible to any volcanism. The special type of volcanic geoheritage elements reflects the conceptual volcano model

framework such as the magma segregation, magma transportation from source to surface, the magma vesiculation, crystallization and fragmentation processes, the eruptive products transportation and deposition modes (either explosive or effusive the process) and the entire set of processes responsible by the remobilization, redeposition and reworking of volcanic material on the surface. Volcanic geoheritage elements can be categorized using general volcano models such as geodynamic settings, monogenetic versus polygenetic nature and chemical compositional distinction as well as the typical volcano architecture and associated volcanic facies. Only by applying complete volcano models to identify volcanic geoheritage elements will lead the correct view of a volcanic terrain be understood and scientifically established. Volcanic geoheritage elements are often viewed as independent from the volcanic region's geocultural and it is suggested they be treated separately to help identify specific volcanic geosites mostly from geotouristic, geoconservation or geoeducation purposes. Geoconservation strategies are also recommended to embrace this concept as this way will guarantee the strong scientifically established backbone of specific conservation strategies. And finally, geoparks can be major and most significant conservation sites where volcanic geoheritage elements and their identified sites should form a logically designed and carefully interlinked array of concepts where we can transmit information to the general audience through various media including modern technologies (e.g., augmented reality, virtual reality, remote sensing, GIS etc.).

Acknowledgements This chapter is a result of many years of work with great scholars within the science of volcanology, geoheritage and geoeducation. The Author acknowledges the support from the fellow members of The Geoconservation Trust Aotearoa Pacific who provided a continuous platform of discussions top develop ideas further. This manuscript is part of outputs associated with the UNESCO IGCP Project 679 "Geoheritage for Geohazard Resilient" initiative. Critical reading and internal review by Julie Palmer are greatly appreciated.

References

Abratis M, Viereck L, Buechner J, Tietz O (2015) Route to the Volcanoes in Germany—conceptual model for a geotourism project interconnecting geosites of Cenozoic volcanism. Zeitschrift Der Deutschen Gesellschaft Fur Geowissenschaften 166(2):161–185

Agustin-Flores J, Nemeth K, Cronin SJ, Lindsay JM, Kereszturi G, Brand BD, Smith IEM (2014) Phreatomagmatic eruptions through unconsolidated coastal plain sequences, Maungataketake, Auckland volcanic field (New Zealand). J Volcanol Geoth Res 276:46–63

Albani RA, Mansur KL, Carvalho IdS, Sa dos Santos WF (2020) Quantitative evaluation of the geosites and geodiversity sites of Joao Dourado Municipality (Bahia-Brazil). Geoheritage 12(2)

Alloway BV, Pillans BJ, Lione CD, Naish TR, Westgate JA (2005) Onshore-offshore correlation of Pleistocene rhyolitic eruptions from New Zealand: implications for TVZ eruptive history and paleoenvironmental construction. Quatern Sci Rev 24(14–15):1601–1622

Antonio Naranjo J, Villa V, Ramirez C, Perez de Arce C (2018) Volcanism and tectonism in the southern Central Andes: tempo, styles, and relationships. Geosphere 14(2):626–641

Araujo AM, Pereira DI (2018) A new methodological contribution for the geodiversity assessment: applicability to Ceara State (Brazil). Geoheritage 10(4):591–605

Argyriou AV, Sarris A, Teeuw RM (2016) Using geoinformatics and geomorphometrics to quantify the geodiversity of Crete, Greece. Int J Appl Earth Obs Geoinf 51:47–59

Balmuth MS, Chester DK, Johnston PA (2005) Cultural responses to the volcanic landscape: the Mediterranean and beyond. In: Balmuth MS, Chester DK, Johnston PA (eds) Archaeological Institute of America, Boston, Mass

Bandopadhyay PC (2005) Discovery of abundant pyroclasts in the Namunagarh Grit, South Andaman: evidence for arc volcanism and active subduction during the Palaeogene in the Andaman area. J Asian Earth Sci 25(1):95–107

Barker SJ, Wilson CJN, Illsley-Kemp F, Leonard GS, Mestel ERH, Mauriohooho K, Charlier BLA (2021) Taupo: an overview of New Zealand's youngest supervolcano. NZ J Geol Geophys 64(2–3):320–346

Becerra-Ramirez R, Gosalvez RU, Escobar E, Gonzalez E, Serrano-Paton M, Guevara D (2020) Characterization and geo-tourist resources of the Campo de Calatrava Volcanic Region (Ciudad Real, Castilla-La Mancha, Spain) to develop a UNESCO global geopark project. Geosciences 10(11)

Beltran-Yanes E, Doniz-Paez J, Esquivel-Sigut I (2020) Chinyero volcanic landscape trail (Canary Islands, Spain): a geotourism proposal to identify natural and cultural heritage in volcanic areas. Geosciences 10(11)

Benedek K, Nagy ZR, Dunkl I, Szabo C, Jozsa S (2001) Petrographical, geochemical and geochronological constraints on igneous clasts and sediments hosted in the Oligo-Miocene Bakony Molasse, Hungary: evidence for a Paleo-Drava River system. Int J Earth Sci 90(3):519–533

Benito-Calvo A, Perez-Gonzalez A, Magri O, Meza P (2009) Assessing regional geodiversity: the Iberian Peninsula. Earth Surf Proc Land 34(10):1433–1445

Bennouna A, Ben Abbou M, Hoepffner C, Kharbouch F, Youbi N (2004) The Carboniferous volcano-sedimentary depocentre of Tazekka Massif (Middle-Atlas, Morocco): new observations and geodynamic implications. J Afr Earth Sci 39(3–5):359–368

Betard F, Peulvast J-P (2019) Geodiversity hotspots: concept, method and cartographic application for geoconservation purposes at a regional scale. Environ Manage 63(6):822–834

Bidias LAZA, Ilouga DCI, Moundi A, Nsangou A (2020) Inventory and assessment of the mbepit massif geomorphosites (cameroon volcanic line): assets for the development of local geotourism. Geoheritage 12(2)

Black BA, Neely RR, Manga M (2015) Campanian Ignimbrite volcanism, climate, and the final decline of the Neanderthals. Geology 43(5):411–414

Bradbury J (2014) A keyed classification of natural geodiversity for land management and nature conservation purposes. Proc Geol Assoc 125(3):329–349

Branney M, Acocella V (2015) Chapter 16: Calderas. In: Sigurdsson H (ed) The encyclopedia of volcanoes, 2nd edn. Academic Press, Amsterdam, pp 299–315

Bratic M, Marjanovic M, Radivojevic AR, Pavlovic M (2020) M-GAM method in function of tourism potential assessment: case study of the Sokobanja basin in eastern Serbia. Open Geosciences 12 (1):1468–1485

Breitkreuz C, de Silva SL, Wilke HG, Pfaender JA, Renno AD (2014) Neogene to quaternary ash deposits in the coastal cordillera in northern Chile: distal ashes from supereruptions in the central Andes. J Volcanol Geoth Res 269:68–82

Brilha J (2016) Inventory and quantitative assessment of geosites and geodiversity sites: a review. Geoheritage 8(2):119–134

Brilha J (2018a) Chapter 4: Geoheritage: inventories and evaluation. In: Reynard E, Brilha J (eds) Geoheritage. Elsevier, pp 69–85

Brilha J (2018b) Chapter 18: Geoheritage and geoparks. In: Reynard E, Brilha J (eds) Geoheritage. Elsevier, pp 323–335

Brocx M, Semeniuk V (2007) Geoheritage and geoconservation—history, definition, scope and scale. J R Soc Western Austr 90(Part 2):53–87

Brocx M, Semeniuk V (2019) The 8Gs'—a blueprint for geoheritage, geoconservation, geo-education and geotourism. Aust J Earth Sci 66 (6):803–821

Brocx M, Semeniuk V (2020) Geodiversity and the '8Gs': a response to Gray & Gordon (2020). Aust J Earth Sci 67(3):445–451

Brocx M, Semeniuk V, Casadevall TJ, Tormey D (2021) Volcanoes: identifying and evaluating their significant geoheritage features from the large to small scale. In: Németh K (ed) Updates in volcanology—transdisciplinary nature of volcano science. IntechOpen, Rijeka, Croatia, pp 329–346. https://doi.org/10.5772/intechopen.97928. Available from https://www.intechopen.com/chapters/76763

Bryan SE, Peate IU, Peate DW, Self S, Jerram DA, Mawby MR, Marsh JS, Miller JA (2010) The largest volcanic eruptions on earth. Earth Sci Rev 102(3–4):207–229

Budai T, Németh K, Piros O (2005) Kozepso-triasz platformkarbonatok es vulkanitok vizsgalata a Latemar kornyeken (Dolomitok, Olaszorszag). Magyar Allami Foldtani Intezet Evi Jelentese = Ann Rep Hungarian Geol Inst 2004:175–188

Cas R, Wright J (1987) Volcanic successions, modern and ancient. Allen and Unwin, London Boston Sydney Wellington, p 528

Cas RAF, Wright JV (1988) Volcanic Successions: modern and ancient —a geological approach to process, products and successions. Chapman and Hall, London, 528 pp

Casadevall TJ, Tormey D, Roberts J (2019) World heritage volcanoes: classification, gap analysis, and recommendations for future listings. IUCN, Gland, Switzerland, p 68

Cashman KV, Cronin SJ (2008) Welcoming a monster to the world: myths, oral tradition, and modern societal response to volcanic disasters. J Volcanol Geoth Res 176(3):407–418

Cashman KV, Giordano G (2008) Volcanoes and human history. J Volcanol Geoth Res 176(3):325–329

Cassinis G, Cortesogno L, Gaggero L, Perotti CR, Buzzi L (2008) Permian to triassic geodynamic and magmatic evolution of the Brescian Prealps (eastern Lombardy, Italy). Boll Soc Geol Ital 127 (3):501–518

Catana MM, Brilha JB (2020) The role of UNESCO global geoparks in promoting geosciences education for sustainability. Geoheritage 12 (1)

Cattell HJ, Cole JW, Oze C (2016) Volcanic and sedimentary facies of the Huka group arc-basin sequence, Wairakei-Tauhara geothermal field, New Zealand. NZ J Geol Geophys 59(2):236–256

Cecioni A, Pineda V (2006) Ethics, geological risks, politics and society. In: Martin-Duque JF, Brebbia CA, Emmanouloudis DE, Mander U (eds) Geo-environment and landscape evolution Ii: evolution, monitoring, simulation, management and remediation of the geological environment and landscape, vol 89, pp 7−+

Challis GA (1960) Igneous rocks in the cape palliser area. NZ J Geol Geophys 3(3):524–542

Christenson B, Németh K, Rouwet D, Tassi F, Vandemeulebrouck J, Varekamp JC (2015) Volcanic lakes. In: Rouwet D, Christenson B, Tassi F, Vandemeulebrouck J (eds) Volcanic lakes. Springer, Berlin, Heidelberg, pp 1–20

Clifford P, Semeniuk V (2019) Sedimentary processes, stratigraphic sequences and middens: the link between archaeology and geoheritage—a case study from the quaternary of the Broome region, Western Australia. Aust J Earth Sci 66(6):955–972

Coratza P, Reynard E, Zwolinski Z (2018) Geodiversity and geoheritage: crossing disciplines and approaches. Geoheritage 10(4):525–526

Cronin SJ, Gaylord DR, Charley D, Alloway BV, Wallez S, Esau JW (2004a) Participatory methods of incorporating scientific with traditional knowledge for volcanic hazard management on Ambae Island, Vanuatu. Bull Volcanol 66(7):652–668

Cronin SJ, Neall VE (2000) Impacts of volcanism an pre-European inhabitants of Taveuni, Fiji. Bull Volcanol 62(3):199–213

Cronin SJ, Petterson MG, Taylor PW, Biliki R (2004b) Maximising multi-stakeholder participation in government and community volcanic hazard management programs; a case study from Savo, Solomon Islands. Nat Hazards 33(1):105–136

Cuevas-Gonzalez J, Diez-Canseco D, Alfaro P, Andreu JM, Baeza-Carratala JF, Benavente D, Blanco-Quintero IF, Canaveras JC, Corbi H, Delgado J, Giannetti A, Martin-Rojas I, Medina I, Peral J, Pla C, Rosa-Cintas S (2020) Geogymkhana-Alicante (Spain): geoheritage through education. Geoheritage 12(1)

Davidson J (2011) Archaeological investigations at Maungarei: a large Maori settlement on a volcanic cone in Auckland, New Zealand. Tuhinga 22:19–100

Davidson JM (1993) The chronology of occupation on Maungarei (Mount Wellington): a large volcanic cone pa in Aucldand. N Z J Archaeol 15:39–55

De Benedetti AA, Funiciello R, Giordano G, Diano G, Caprilli E, Paterne M (2008) Volcanology, history and myths of the Lake Albano maar (Colli Albani volcano, Italy). J Volcanol Geoth Res 176(3):387–406

de Silva S, Lindsay JM (2015) Chapter 15: Primary volcanic landforms. In: Sigurdsson H (ed) The encyclopedia of volcanoes, 2nd edn. Academic Press, Amsterdam, pp 273–297

De Vries BVW, Byrne P, Delcamp A, Einarson P, Gogus O, Guilbaud M-N, Hagos M, Harangi S, Jerram D, Matenco L, Mossoux S, Németh K, Maghsoudi M, Petronis MS, Rapprich V, Rose WI, Vye E (2018) A global framework for the earth: putting geological sciences in context. Glob Planet Change 171:293–321

Dias MCSS, Domingos JO, dos Santos Costa SS, do Nascimento MAL, da Silva MLN, Granjeiro LP, de Lima Miranda RF (2021) Geodiversity index map of Rio Grande do Norte State, Northeast Brazil: cartography and quantitative assessment. Geoheritage 13(1)

Dóniz-Páez J, Becerra-Ramirez R, Gonzalez-Cardenas E, Guillen-Martin C, Escobar-Lahoz E (2011) Geomorphosites and geotourism in volcanic landscape: the example of La Corona del Lajial cinder cone (El Hierro, Canary Islands, Spain). GeoJ Tourism Geosites 8(2):185–197

Dóniz-Páez J, Beltran-Yanes E, Becerra-Ramirez R, Perez NM, Hernandez PA, Hernandez W (2020) Diversity of volcanic geoheritage in the Canary Islands, Spain. Geosciences 10(10)

Donovan K (2010) Doing social volcanology: exploring volcanic culture in Indonesia. Area 42(1):117–126

Dunkl I, Farics E, Jozsa S, Lukacs B, Haass J, Budai T (2019) Traces of Carnian volcanic activity in the Transdanubian Range, Hungary. Int J Earth Sci 108(5):1451–1466

Erfurt-Cooper P (2011) Geotourism in volcanic and geothermal environments: playing with fire? Geoheritage 3(3):187–193

Erfurt-Cooper P (2014) Volcanic tourist destinations. In: Eder W, Bobrowsky PT, Martínez-Frías J (eds) Geoheritage, geoparks and geotourism. Springer, Heidelberg, p 384

Erfurt P (2018) Chapter 11: Geotourism development and management in volcanic regions. In: Dowling R, Newsome D (eds) Handbook of geotourism. Edward Elgar Publishing, Cheltenham Glos, UK, pp 152–167. ISBN: 978-1-78536-885-1

Escorihuela J (2018) The role of the geotouristic guide in earth science education: towards a more critical society of land management. Geoheritage 10(2):301–310

Fepuleai A, Nemeth K (2019) Volcanic geoheritage of landslides and rockfalls on a tropical ocean island (Western Samoa, SW Pacific). Geoheritage 11(2):577–596

Fepuleai A, Németh K, Muliaina T (2021) Geopark Impact for the resilience of communities in Samoa, SW Pacific. Geoheritage 13(3)

Fepuleai A, Weber E, Nemeth K, Muliaina T, Iese V (2017) Eruption styles of samoan volcanoes represented in tattooing, language and cultural activities of the indigenous people. Geoheritage 9(3):395–411

Floyd JD (2001) The southern uplands terrane: a stratigraphical review. Trans R Soc Edinburgh-Earth Sci 91:349–362

Fox A (1983) Pa and people in New Zealand: an archaeological estimate of population. N Z J Archaeol 5:5–18

Fox N, Graham LJ, Eigenbrod F, Bullock JM, Parks KE (2020) Incorporating geodiversity in ecosystem service decisions. Ecosyst People 16(1):151–159

Furrer H, Schaltegger U, Ovtcharova M, Meister P (2008) U-Pb zircon age of volcaniclastic layers in middle triassic platform carbonates of the Austroalpine Silvretta nappe (Switzerland). Swiss J Geosci 101 (3):595–603

Gamberi F (2001) Volcanic facies associations in a modern volcaniclastic apron (Lipari and Vulcano offshore, Aeolian Island Arc). Bull Volcanol 63(4):264–273

Gordon JE, Barron HF (2013) The role of geodiversity in delivering ecosystem services and benefits in Scotland. Scott J Geol 49:41–58

Graettinger AH (2018) Trends in maar crater size and shape using the global Maar Volcano location and shape (MaarVLS) database. J Volcanol Geoth Res 357:1–13

Gravis I, Nemeth K, Procter JN (2017) The role of cultural and indigenous values in geosite evaluations on a quaternary monogenetic volcanic landscape at IhumAtao, Auckland volcanic field, New Zealand. Geoheritage 9(3):373–393

Gravis I, Németh K, Twemlow C, Németh B (2020) The case for community-led geoheritage and geoconservation ventures in Mangere, South Auckland, and Central Otago, New Zealand. Geoheritage 12(1)

Gray M (2012) Valuing geodiversity in an 'ecosystem services' context. Scottish Geogr J 128(3–4):177–194

Gray M (2018a) Chapter 1: Geodiversity: the backbone of geoheritage and geoconservation. In: Reynard E, Brilha J (eds) Geoheritage. Elsevier, pp 13–25

Gray M (2018b) Chapter 3: Geodiversity, geoheritage, geoconservation and their relationship to geotourism. In: Dowling R, Newsome D (eds) Handbook of geotourism. Edward Elgar Publishing, Cheltenham Glos, UK, pp 48–60. ISBN: 978-1-78536-885-1

Gray M (2018c) The confused position of the geosciences within the "natural capital" and "ecosystem services" approaches. Ecosyst Serv 34:106–112

Gray M, Gordon JE (2020) Geodiversity and the '8Gs': a response to Brocx & Semeniuk (2019). Aust J Earth Sci 67(3):437–444

Gruber G (2007) The Messel maar. Hessisches Landesmuseum Darmstadt, Germany (DEU)

Gudmundsson MT (2015) Chapter 56: Hazards from Lahars and Jökulhlaups. In: Sigurdsson H (ed) The encyclopedia of volcanoes, 2nd edn. Academic Press, Amsterdam, pp 971–984

Guilbaud M-N, del Pilar Ortega-Larrocea M, Cram S, de Vries BvW (2021) Xitle volcano geoheritage, Mexico City: raising awareness of natural hazards and environmental sustainability in active volcanic areas. Geoheritage 13(1)

Harangi S (2014) Volcanic heritage of the Carpathian-Pannonian Region in Eastern-Central Europe. In: Erfurt-Cooper P (ed) Volcanic tourist destinations. Springer, Berlin Heidelberg, Berlin, Heidelberg, pp 103–123

Henriques MH, Brilha J (2017) UNESCO global geoparks: a strategy towards global understanding and sustainability. Episodes 40 (4):349–355

Henriques MH, dos Reis RP, Brilha J, Mota T (2011) Geoconservation as an emerging geoscience. Geoheritage:1–12

Hlusek R (2020) Ritual landscape and sacred mountains in past and present Mesoamerica. Rev Anthropol 49(1–2):39–60

Hodgson KA, Nairn IA (2005) The c. AD 1315 syn-eruption and AD 1904 post-eruption breakout floods from Lake Tarawera, Haroharo caldera, North Island, New Zealand. N Z J Geol Geophys 48 (3):491–506

Hopkins JL, Smid ER, Eccles JD, Hayes JL, Hayward BW, McGee LE, van Wijk K, Wilson TM, Cronin SJ, Leonard GS, Lindsay JM, Nemeth K, Smith IEM (2021) Auckland volcanic field magmatism, volcanism, and hazard: a review. NZ J Geol Geophys 64(2–3):213–234

Ibanez K, Garcia MdGM, Mazoca CEM (2021) Tectonic geoheritage as records of western gondwana history: a study based on a geosite's potential in the Central Ribeira Belt, Southeastern Brazil. Geoheritage 13(1)

Kelman I, Mather TA (2008) Living with volcanoes: the sustainable livelihoods approach for volcano-related opportunities. J Volcanol Geoth Res 172(3–4):189–198

Kereszt:uri G, Bebbington M, Nemeth K (2017) Forecasting transitions in monogenetic eruptions using the geologic record. Geology 45 (3):283–286

Kereszt*uri G, Nemeth K (2016) Sedimentology, eruptive mechanism and facies architecture of basaltic scoria cones from the Auckland volcanic field (New Zealand). J Volcanol Geoth Res 324:41–56

Kereszt#uri G, Németh K, Cronin SJ, Procter J, Agustin-Flores J (2014) Influences on the variability of eruption sequences and style transitions in the Auckland volcanic field, New Zealand. J Volcanol Geoth Res 286:101–115

Kilburn CRJ (2000) Lava flows and flow fields. In: Sigurdsson H, Houghton BF, McNutt SR, Rymer H, Stix J (eds) Encyclopedia of volcanoes. Academic Press, San Diego, pp 291–306

Kovács J, Németh K, Szabó P, Kocsis L, Keresztúri G, Újvári G, Vennemann T (2020) Volcanism and paleoenvironment of the Pula maar complex: a pliocene terrestrial fossil site in Central Europe (Hungary). Palaeogeogr Palaeoclimatol Palaeoecol 537:15

Kubalikova L (2020) Cultural ecosystem services of geodiversity: a case study from Stranska skala (Brno, Czech Republic). Land 9(4)

Latutrie B, Ross P-S (2019) Transition zone between the upper diatreme and lower diatreme: origin and significance at Round Butte, Hopi Buttes volcanic field, Navajo Nation, Arizona. Bull Volcanol 81(4)

Laumonier B (1998) Central and Eastern Pyrenees at the beginning of the Paleozoic (Cambrian sl): Paleogeographic and geodynamic evolution. Geodin Acta 11(1):1–11

Lefebvre NS, White JDL, Kjarsgaard BA (2013) Unbedded diatreme deposits reveal maar-diatreme-forming eruptive processes: standing rocks west, Hopi Buttes, Navajo Nation, USA. Bull Volcanol 75(8)

Lenz OK, Wilde V (2018) Changes in eocene plant diversity and composition of vegetation: the lacustrine archive of Messel (Germany). Paleobiology 44(4):709–735

Lewis ID (2020) Linking geoheritage sites: geotourism and a prospective geotrail in the flinders ranges world heritage nomination area, South Australia. Aust J Earth Sci 67(8):1195–1210

Lim K (2014) A study of geotourism growth through recognition of geoeducation and geoconservation for the geoheritage. J Tour Leisure Res 26(6):43–59

Lindsay JM, de Silva S, Trumbull R, Emmermann R, Wemmer K (2001) La Pacana caldera, N. Chile: a re-evaluation of the stratigraphy and volcanology of one of the world's largest resurgent calderas. J Volcanol Geothermal Res 106(1–2):145–173

Lorenz V (2007) Syn- and posteruptive hazards of maar-diatreme volcanoes. J Volcanol Geothermal Res 159(1–3):285–312

Lowe DR, Williams SN, Leigh H, Connor CB, Gemmell JB, Stoiber RE (1986) Lahars initiated by the 13 November 1985 eruption of Nevado-del-Ruiz, Colombia. Nature 324(6092):51–53

Macadam J (2018) Chapter 15: Geoheritage: getting the message across. What message and to whom? In: Reynard E, Brilha J (eds) Geoheritage. Elsevier, pp 267–288

Manville V (2002) Sedimentary and geomorphic responses to ign-imbrite emplacement: readjustment of the Waikato River after the AD 181 Taupo Eruption, New Zealand. J Geol 110(5):519–541

Manville V, Hodgson KA, Nairn IA (2007) A review of break-out floods from volcanogenic lakes in New Zealand. N Z J Geol Geophys 50(2):131–150

Marin A, Vergara-Pinto F, Prado F, Farias C (2020) Living near volcanoes: scoping the gaps between the local community and volcanic experts in Southern Chile. J Volcanol Geothermal Res 398

Martí J, Groppelli G, Brum da Silveira A (2018) Volcanic stratigraphy: a review. J Volcanol Geoth Res 357:68–91

Megerle HE (2020a) Geoheritage and geotourism in regions with extinct volcanism in Germany; case study Southwest Germany with UNESCO global geopark Swabian Alb. Geosciences 10(11):445

Megerle HE (2020b) Geoheritage and geotourism in regions with extinct volcanism in Germany; case study Southwest Germany with UNESCO global geopark Swabian Alb. Geosciences 10(11)

Megerssa L, Rapprich V, Novotny R, Verner K, Erban V, Legesse F, Manaye M (2019) Inventory of key geosites in the Butajira volcanic field: perspective for the first geopark in Ethiopia. Geoheritage 11 (4):1643–1653

Michaux B, Ebach MC, Dowding EM (2018) Cladistic methods as a tool for terrane analysis: a New Zealand example. N Z J Geol Geophys 61(2):127–135

Migon P, Pijet-Migon E (2016) Overlooked geomorphological component of volcanic geoheritage-diversity and perspectives for tourism industry, Pogrze Kaczawskie Region, SW Poland. Geoheritage 8(4):333–350

Migon P, Pijet-Migon E (2020) Late palaeozoic volcanism in Central Europe—geoheritage significance and use in geotourism. Geoheritage 12(2)

Moufti MR, Nemeth K, El-Masry N, Qaddah A (2013a) Geoheritage values of one of the largest maar craters in the Arabian Peninsula: the Al Wahbah Crater and other volcanoes (Harrat Kishb, Saudi Arabia). Central Eur J Geosci 5(2):254–271

Moufti MR, Nemeth K, El-Masry N, Qaddah A (2015) Volcanic geotopes and their geosites preserved in an arid climate related to landscape and climate changes since the neogene in Northern Saudi Arabia: Harrat Hutaymah (Hai'il Region). Geoheritage 7(2):103–118

Moufti MR, Németh K, El-Masry N, Qaddah A (2013b) Geoheritage values of one of the largest maar craters in the Arabian Peninsula: the Al Wahbah Crater and other volcanoes (Harrat Kishb, Saudi Arabia). Central Euro J Geosci 5(2):254–271

Murcia HF, Hurtado BO, Cortes GP, Macias JL, Cepeda H (2008) The similar to 2500 yr BP chicoral non-cohesive debris flow from Cerro Machin Volcano, Colombia. J Volcanol Geoth Res 171(3–4):201–214

Nahuelhual L, Benra Ochoa F, Rojas F, Ignacio Diaz G, Carmona A (2016) Mapping social values of ecosystem services: what is behind the map? Ecol Soc 21(3)

Németh B, Németh K, Procter JN (2021a) Informed geoheritage conservation: determinant analysis based on bibliometric and sustainability indicators using ordination techniques. Land 10(5)

Németh B, Németh K, Procter JN, Farrelly T (2021) Geoheritage conservation: systematic mapping study for conceptual synthesis. Geoheritage 13(2):45

Németh K (2016a) Geoheritage values of monogenetic volcanic fields: a potential UNESCO world heritage site in the Arabian Peninsula. In: 6th international maar conference. inTech Open Publisher, Changchun, China, Rijeka, Croatia, pp 67–68

Nemeth K (2016b) Volcanic geoheritage values of monogenetic volcanic fields in the global scale and from New Zealand perspective. Geosci Soc N Z Miscellaneous Publ 145A:59

Németh K (2017) Pedagogical and indigenous aspects of geoeducation with special reference to the volcanic geoheritage of monogenetic volcanoes. In: IAVCEI 2017 Scientific Assembly. Portland, Oregon

Németh K, Budai T (2009) Diatremes cut through the triassic carbonate platforms in the dolomites? Evidences from and around the Latemar, northern Italy. Episodes 32(2):74–83

Nemeth K, Casadevall T, Moufti MR, Marti J (2017) Volcanic geoheritage. Geoheritage 9(3):251–254

Németh K, Cronin SJ (2009) Volcanic structures and oral traditions of volcanism of Western Samoa (SW Pacific) and their implications for hazard education. J Volcanol Geoth Res 186(3–4):223–237

Németh K, Cronin SJ, Stewart RB, Charley D (2009) Intra- and extra-caldera volcaniclastic facies and geomorphic characteristics of a frequently active mafic island-arc volcano, Ambrym Island, Vanuatu. Sedimentary Geol 220(3–4):256–270

Nemeth K, Goth K, Martin U, Csillag G, Suhr P (2008) Reconstructing paleoenvironment, eruption mechanism and paleomorphology of the Pliocene Pula maar, (Hungary). J Volcanol Geoth Res 177 (2):441–456

Németh K, Gravis I, Németh B (2021b) Dilemma of geoconservation of monogenetic volcanic sites under fast urbanization and infrastructure developments with special relevance to the Auckland volcanic field, New Zealand. Sustainability 13(12)

Nemeth K, Kereszturi G (2015) Monogenetic volcanism: personal views and discussion. Int J Earth Sci 104(8):2131–2146

Németh K, Palmer J (2019) Geological mapping of volcanic terrains: discussion on concepts, facies models, scales, and resolutions from New Zealand perspective. J Volcanol Geoth Res 385:27–45

Newhall CG, Self S (1982) The volcanic explosivity index (VEI) an estimate of explosive magnitude for historical volcanism. J Geophys Res Oceans 87(C2):1231–1238

Nunn PD, Baniala M, Harrison M, Geraghty P (2006) Vanished islands in Vanuatu: new research and a preliminary geohazard assessment. J R Soc N Z 36(1):37–50

Nunn PD, Lancini L, Franks L, Compatangelo-Soussignan R, McCallum A (2019) Maar stories: how oral traditions aid understanding of maar volcanism and associated phenomena during preliterate times. Ann Am Assoc Geogr 109(5):1618–1631

Oppenheimer C, Orchard A, Stoffel M, Newfield TP, Guillet S, Corona C, Sigl M, Di Cosmo N, Buntgen U (2018) The Eldgja eruption: timing, long-range impacts and influence on the christianisation of Iceland. Clim Change 147(3–4):369–381

Pal M, Albert G (2021) Examining the spatial variability of geosite assessment and its relevance in geosite management. Geoheritage 13(1)

Pardo N, Espinosa ML, Gonzalez-Arango C, Cabrera MA, Salazar S, Archila S, Palacios N, Prieto D, Camacho R, Parra-Agudelo L (2021) Worlding resilience in the Dona Juana Volcano-Paramo, Northern Andes (Colombia): a transdisciplinary view. Natural Hazards

Pardo N, Wilson H, Procter JN, Lattughi E, Black T (2015) Bridging Māori indigenous knowledge and western geosciences to reduce social vulnerability in active volcanic regions. J Appl Volcanol 4 (1):5

Paulo A, Galas A, Galas S (2014) Planning the Colca Canyon and the Valley of the Volcanoes National Park in South Peru. Environ Earth Sci 71(3):1021–1032

Petterson MG (2019) Interconnected geoscience for international development. Episodes 42(3):225–233

Petterson MG, Cronin SJ, Taylor PW, Tolia D, Papabatu A, Toba T, Qopoto C (2003) The eruptive history and volcanic hazards of Savo, Solomon Islands. Bull Volcanol 65(2–3):165–181

Pijet-Migon E (2016) Geotourism—new opportunities to use geoheritage for tourism development. Case study of land of extinct volcanoes in the West Sudetes. Ekonomiczne Problemy Turystyki 1:301–311

Pijet-Migon E, Migon P (2019) Promoting and interpreting geoheritage at the local level-bottom-up approach in the land of extinct Volcanoes, Sudetes, SW Poland. Geoheritage 11(4):1227–1236

Pillans B, Alloway B, Naish T, Westgate J, Abbott S, Palmer A (2005) Silicic tephras in Pleistocene shallow-marine sediments of Wanganui Basin, New Zealand. J R Soc N Z 35(1–2):43–90

Planaguma L, Marti J (2020) Identification, cataloguing and preservation of outcrops of geological interest in monogenetic volcanic fields: the case of La Garrotxa Volcanic Zone Natural Park. Geoheritage 12(4)

Plunket P, Urunuela G (2005) Recent research in Puebla prehistory. J Archaeol Res 13(2):89–127

Procter J, Zernack A, Mead S, Morgan M, Cronin S (2021) A review of lahars; past deposits, historic events and present-day simulations from Mt. Ruapehu and Mt. Taranaki, New Zealand. N Z J Geol Geophys 64(2–3):479–503

Quesada-Román A, Pérez-Umaña D (2020) State of the art of geodiversity, geoconservation, and geotourism in Costa Rica. Geosciences 10(6):211

Rapprich V, Lisec M, Fiferna P, Zavada P (2017) Application of modern technologies in popularization of the Czech volcanic geoheritage. Geoheritage 9(3):413–420

Rees C, Palmer A, Palmer J (2019) Quaternary sedimentology and tephrostratigraphy of the lower Pohangina Valley, New Zealand. N Z J Geol Geophys 62(2):171–194

Rees C, Palmer J, Palmer A (2018) Plio-Pleistocene geology of the lower Pohangina Valley, New Zealand. NZ J Geol Geophys 61 (1):44–63

Rees C, Palmer J, Palmer A (2020) Tephrostratigraphic constraints on sedimentation and tectonism in the Whanganui Basin, New Zealand. N Z J Geol Geophys 63(2):262–280

Reynard E, Giusti C (2018) Chapter 8: The landscape and the cultural value of geoheritage. In: Reynard E, Brilha J (eds) Geoheritage. Elsevier, pp 147–166

Riguccio L, Russo P, Scandurra G, Tomaselli G (2015) Cultural landscape: stone towers on mount etna. Landsc Res 40(3):294–317

Roser B, Grapes R (1990) Geochemistry of a metabasite chert colored-argillite turbidite association at red rocks, Wellington, New Zealand. N Z J Geol Geophys 33(2):181–191

Ruban DA (2017) Geodiversity as a precious national resource: a note on the role of geoparks. Resour Policy 53:103–108

Scarlett JP, Riede F (2019) The dark geocultural heritage of volcanoes: combining cultural and geoheritage perspectives for mutual benefit. Geoheritage 11(4):1705–1721

Schwartz-Marin E, Merli C, Rachmawati L, Horwell C, Nugroho F (2020) Merapi multiple: protection around Yogyakarta's celebrity volcano through masks, dreams, and seismographs. History Anthropol

Sheth H (2018) Flood basalt landscapes. In: A photographic atlas of flood basalt volcanism. Springer, pp 7–31

Smith IEM, Németh K (2017) Source to surface model of monogenetic volcanism: a critical review In: Németh K, Carrasco-Nuñez G,

Aranda-Gomez JJ, Smith IEM (eds) Monogenetic volcanism. The Geological Society Publishing House, Bath, UK, Geological Society of London, Special Publications, vol 446, pp 1–28

Streeter R, Dugmore AJ, Vesteinsson O (2012) Plague and landscape resilience in premodern Iceland. Proc Natl Acad Sci USA 109 (10):3664–3669

Swanson DA (2008) Hawaiian oral tradition describes 400 years of volcanic activity at Kilauea. J Volcanol Geoth Res 176(3):427–431

Szepesi J, Esik Z, Soos I, Nemeth B, Suto L, Novak TJ, Harangi S, Lukacs R (2020) Identification of geoheritage elements in a cultural landscape: a case study from Tokaj Mts, Hungary. Geoheritage 12(4)

Szepesi J, Harangi S, Esik Z, Novak TJ, Lukacs R, Soos I (2017) Volcanic geoheritage and geotourism perspectives in Hungary: a case of an UNESCO world heritage site, Tokaj Wine region historic cultural landscape, Hungary. Geoheritage 9(3):329–349

Thouret JC, Ramirez JC, Gibert-Malengreau B, Vargas CA, Naranjo JL, Vandemeulebrouck J, Valla F, Funk M (2007) Volcano-glacier interactions on composite cones and lahar generation: Nevado del Ruiz, Colombia, case study. In: Clarke GKC, Smellie J (eds) Annals of glaciology, Vol 45, pp 115−+

Turner S (2013) Geoheritage and geoparks: one (Australian) woman's point of view. Geoheritage 5(4):249–264

Vallance JW (2000) Lahars. In: Sigurdsson H, Houghton BF, McNutt SR, Rymer H, Stix J (eds) Encyclopedia of volcanoes. Academic Press, San Diego, pp 601–616

Vizuete DDC, Velasquez CRC, Marcu MV, Montoya AVG (2020) The use of cultural ecosystem services: a comparison between locals and tourists in the Chimborazo natural reserve. Bull Transilvania Univ Brasov, Series II For Wood Indus Agric Eng 13(62 Part 2):1–18

Voight B (1990) The 1985 Nevado-del-Ruiz volcano catastrophe—anatomy and retrospection. J Volcanol Geoth Res 44(3–4):349–386

Vörös F, Pál M, van Wyk de Vries B, Székely B (2021) Development of a new type of geodiversity system for the Scoria Cones of the Chaîne des Puys based on geomorphometric studies. Geosciences 11(2):58

Vujicic MD, Vasiljevic DA, Markovic SB, Hose TA, Lukic T, Hadzic O, Janicevic S (2011) Preliminary geosite assessment model (GAM) and its application on Fruska Gora Mountain, potential geotourism destination of Serbia. Acta Geographica Slovenica-Geografski Zbornik 51(2):361–376

Walker GPL (1973) Explosive volcanic eruptions—a new classification scheme. Geologische Rundschau 62(2):431–446

Wilkie B, Cahir F, Clark ID (2020) Volcanism in aboriginal Australian oral traditions: ethnographic evidence from the newer volcanics province. J Volcanol Geothermal Res 403

Wolniewicz P (2021) Beyond geodiversity sites: exploring the educational potential of widespread geological features (rocks, minerals and fossils). Geoheritage 13(2)

Wood C (2009) World heritage volcanoes: a thematic study. A global review of volcanic world heritage properties: present situation, future prospects and management requirements international union for conservation of nature and natural resources, Gland, Switzerland

Yepez Noboa AM (2020) Empty spaces that are full of cultural history: an innovative proposal for the management of a protected area of Chimborazo volcano (Ecuaclor). Eco Mont-J Protected Mountain Areas Res 12(1):43–49

Yoon SH (2019) Geoheritage value and utilization plan of the Hanon volcanic crater, Jeju Island. J Geol Soc Korea 55(3):353–363

Zangmo GT, Kagou AD, Nkouathio DG, Gountie MD, Kamgang P (2017) The volcanic geoheritage of the mount Bamenda Calderas (cameroon line): assessment for geotouristic and geoeducational purposes. Geoheritage 9(3):255–278

Zeidler JA (2016) Modeling cultural responses to volcanic disaster in the ancient Jama-Coaque tradition, coastal Ecuador: a case study in cultural collapse and social resilience. Quatern Int 394:79–97

Zemeny A, Procter J, Németh K, Zellmer GF, Zernack AV, Cronin SJ (2021) Elucidating stratovolcano construction from volcaniclastic mass-flow deposits: the medial ring-plain of Taranaki Volcano, New Zealand. Sedimentology. https://doi.org/10.1111/sed.12857

Zernack AV (2021) Volcanic debris-avalanche deposits in the context of volcaniclastic ring plain successions—a case study from Mt. Taranaki. In: Roverato M, Dufresne A, Procter J (eds) Volcanic debris avalanches: from collapse to hazard. Springer International Publishing, Cham, pp 211–254

Zernack AV, Cronin SJ, Neall VE, Procter JN (2011) A medial to distal volcaniclastic record of an andesite stratovolcano: detailed stratigraphy of the ring-plain succession of south-west Taranaki, New Zealand. Int J Earth Sci 100(8):1937–1966

Zernack AV, Procter JN (2021) Cyclic growth and destruction of volcanoes. In: Roverato M, Dufresne A, Procter J (eds) Volcanic debris avalanches: from collapse to hazard. Springer International Publishing, Cham, pp 311–355

Zernack AV, Procter JN, Cronin SJ (2009) Sedimentary signatures of cyclic growth and destruction of stratovolcanoes: a case study from Mt. Taranaki, New Zealand. Sedimentary Geol 220(3–4):288–305

Zolitschka B, Anselmetti F, Ariztegui D, Corbella H, Francus P, Luecke A, Maidana NI, Ohlendorf C, Schaebitz F, Wastegard S (2013) Environment and climate of the last 51,000 years—new insights from the Potrok Aike maar lake sediment archive drilling project (PASADO). Quatern Sci Rev 71:1–12

Zwoliński Z, Najwer A, Giardino M (2018) Chapter 2: Methods for assessing geodiversity. In: Reynard E, Brilha J (eds) Geoheritage. Elsevier, pp 27–52

Volcanology of Recent Oceanic Active Island

William Hernández Ramos, Victor Ortega, Monika Przeor,
Nemesio M. Pérez, and Pedro A. Hernández

Abstract

The island of El Hierro is the youngest of the entire Canary archipelago, with an age of about 1.56 My. However, it has had a rapid growth, which has caused that from its first stages of formation it has had important collapses. Since submarine volcanism, El Hierro has gone through different phases of formation such as the construction of the Tiñor Building, later that of the El Golfo Building, then came the Rifts volcanism and finally the historical volcanism. This is the geological context of an island whose formation process has not yet finished.

Keywords

Geology • Formation • El Hierro • Canary Islands

1 Formation and Large Units of the Relief Island

The Canary Islands originated more than 70 My ago from intraplate volcanism in the African Plate (Anguita and Hernán 2000). For decades their formation has caused numerous controversies on which the scientific community has not fully agreed. In general terms, they can be divided into two types of ideas: thermal and tectonic. The thermal ones are related to a mantle plume, also called "hot spot" (Anguita and Hernán 2000). The second idea states that tectonics plays a major role in the origin of the islands. These theories are the propagating fracture and uplifted block theories.

The hot spot or mantle plume theory extrapolates the model of creation of Hawaiian volcanism to Canary volcanism, that is, a magmatic flow that is injected from the mantle to the crust, forming the islands in the vertical of this focus (Sandoval-Velasquez et al. 2021). Among the thermal theories is the theory of the blob or "bubble" model, which is based on a series of magma droplets or bubbles that underlie the archipelago, which are injected into the lithosphere to produce volcanic events. This theory would explain the magmatic cycles that have occurred in the Canary Islands, as well as the geochemical diversity, which would have a simple explanation as a consequence of the heterogeneity of these bubbles (Anguita and Hernán 2000).

On the other hand, the propagating fracture theory links the formation of the islands to the Moroccan Atlas Mountains through a shear fault. The uplifted block theory argues that compressional tectonics led to crustal thickening, causing the uplift of the blocks that formed the islands (Sandoval-Velasquez et al. 2021). Finally, the unified model tried to integrate tectonic and thermal theories to explain the complexity of the formation of the Canary Islands (Sandoval-Velasquez et al. 2021).

El Hierro is the youngest island of the whole Archipelago, with an age of about 1.56 My, which corresponds to submarine construction, although the oldest subaerial rocks have been dated at 1.12 My (Gee et al. 2001). The formation of the island is considered to have occurred rapidly due to the intense processes that have taken place (Gómez Sáinz de Aja et al. 2010). One of the singularities of El Hierro is the clarity of massive lateral landslides and rift volcanism. These facts have resulted in a star-shaped island with structural

W. H. Ramos (✉) · V. Ortega · N. M. Pérez · P. A. Hernández
Volcanological Institute of the Canary Islands (INVOLCAN), Granadilla de Abona, Spain
e-mail: william.hernandez@involcan.org

V. Ortega
e-mail: victor.ortega@involcan.org

N. M. Pérez
e-mail: nperez@iter.es

P. A. Hernández
e-mail: phdez@iter.es

M. Przeor · N. M. Pérez · P. A. Hernández
Institute of Technology and Renewable Energy (ITER), Granadilla de Abona, Spain
e-mail: mprzeor@iter.es

J. Dóniz-Páez and N. M. Pérez (eds.), *El Hierro Island Global Geopark*,
Geoheritage, Geoparks and Geotourism, https://doi.org/10.1007/978-3-031-07289-5_2

axes every 120°. The first volcanic landmark was the Tiñor volcano, which was active between 1.2 and 0.88 My ago (Guillou et al. 1996; Barrera Morate and García Moral 2011). The rapidity of its formation produced instabilities in the platform that led to gravity sliding (Gómez Sáinz de Aja et al. 2010).

This was followed by a significant period of inactivity and then, between 0.54 and 0.17 My, a second stage of formation began in the Tiñor slide basin, corresponding to the eruptions of the El Golfo volcanic edifice. The materials ejected by this volcano completely buried the Tiñor landslide scar and a large part of the preceding volcanic edifice. At the end of this new stage, there were two volcanoes separated by an almost vertical landslide scarp (Carracedo et al. 2001).

Subsequently, rift volcanism began, a stage characterized by several events. The emission centers are grouped in the main structural axes of the island, which are more concentrated in the center and south, and are more dispersed on the east and west flanks. Between 0.5 and 0.3 My was the El Julan gravity slip (Carracedo et al. 2001), of which there is no evidence on the surface, but there is evidence on the ocean floor. Between 0.54 and 0.17 My ago, a new lateral collapse took place to the northeast, at San Andrés (Day et al. 1997), which was not completed and ended up generating a system of step faults. Next, between 0.17 and 0.14 My was the Las Playas gravity slide (Gee et al. 2001), which has the smallest volume of all. The largest landslide occurred at El Golfo, which is also the best example of a large-scale collapse (Gee et al. 2001). However, authors do not agree on the age, although some studies suggest that it could be between 0.013 and 0.017 My (Carracedo et al. 2001). After this mega-sliding, and also within the rift volcanism, there followed a stage of island formation characterized by recent volcanism that took place at various points in the El Golfo valley and filled in the escarpments of this valley. These episodes shaped the geography of the island as it is known today.

The last stage is limited to historical volcanism, represented by the eruption of the Tagoro volcano between 2011 and 2012. This is a submarine eruption that occurred SW of La Restinga and remained at just 88 m (Pérez-Torrado et al. 2012) from sea level. Therefore, volcanism on the island is still active. Both the ages of the different episodes of its formation, as well as the last volcanic landmark, suggest that the activity will continue in the future (Fig. 1).

1.1 Building Tiñor (1.2–0.88 My)

The El Tiñor volcanic edifice, formed during 0.3 My (Gómez Sáinz de Aja et al. 2010), is the first phase of subaerial formation on the island of El Hierro. Thanks to K–Ar and magnetostratigraphic dating methods, its formation is estimated in the Lower-Middle Pleistocene (Barrera Morate

and García Moral 2011). The Tiñor volcanic complex developed very rapidly, with three phases of formation with three well-differentiated phases of formation (Barrera Morate and García Moral 2011). Finally, the large volumes of material emitted and its great height, caused a destabilization of the flank and, consequently, a large gravitational slide exposing the first phases of formation of the Tiñor.

The first stage took place 1.12 My ago with the shield construction of the Tiñor edifice, where volcanic material covered large extensions (Guillou et al. 1996; Carracedo et al. 2008). It was represented by basaltic lava flows of not very great power, with brecciated appearance and divergent dikes of plagioclase basalts and pyroclastic deposits along the eruptive fissures (Barrera Morate and García Moral 2011). The first stage of formation outcrops at the bottom of the Tiñor, Honduras, Balón and Playecillas ravines. The main outcrops are in the NE sector of the Tiñor and in the Las Playas escarpment (lower-middle part). Narrow subvertical dykes intercalate thin lava flows with pyroclasts and brecciated levels. Several buried cones and pyroclast levels can be seen intercalated with the lavas (Barrera Morate and García Moral 2011). They are mainly basaltic, basaltic and tephritic flows. The basalts are plagioclastic-olivine-pyroxenic (Gómez Sáinz de Aja et al. 2010). One of the most abundant geomorphological units, but because they are submerged and one of the least common, are the pillow lavas (pillow-lavas) that can be seen in Timijiraque Bay. Another unit present in the lower section of Tiñor is formed by tephra cones and some buried cones with hydromagmatic and strombolian phases (Barrera Morate and García Moral 2011).

The second stage of formation of the Tiñor edifice occupies larger volumes than the previous stage. It is the intermediate or tabular stage that formed the San Andrés plateau thanks to lavas of great power and with subhorizontal dip in the center of the same building in formation (Guillou et al. 1996; Carracedo et al. 2008). This plateau is dated to between approximately 1.04 and 1.7 My (Barrera Morate and García Moral 2011). The San Andrés plateau formation section outcrops both in the El Toril escarpment and on the Dar slopes, as well as in the Tiñor and Honduras ravines. The materials of the tabular section outcrop in Tamaduste and Puerto de La Estaca and are mainly basaltic, basanitic and tephritic lava flows (Gómez Sáinz de Aja et al. 2010). The compositional typology of the materials includes pyroxenic-olivine basalts, olivine-pyroxenic basalts, olivine basalts, trachybasalts, tephras and plagioclase basalts. The "aa" lavas from these eruptions create a stepped relief with slag intercalations. Within the plateau unit, two basic intrusive bodies are recognized (Barrera Morate and García Moral 2011).

After the end of the intermediate growth stage, between 1.04 and 0.88 My, there was an eruptive lull that culminated

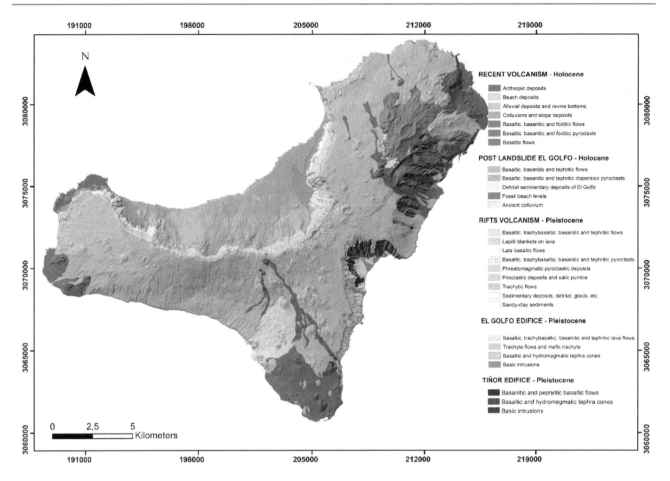

Fig. 1 Geological mapa of El Hierro Island. *Source* Grafcan, self-elaboration

with the last stage of formation of the volcanic complex (Barrera Morate and García Moral 2011). After this eruptive pause, the formation of the Ventejís section took place, known as the late period, with a more explosive character. Wide craters were formed, pyroclasts were deposited interspersed with flows with pyroxene nodules, giving rise to the presence of the Ventejís stratovolcano with a predominant direction of growth towards NE. The formative stage is shown by the alignment of the Picos-Rivera-Moles buildings. The main material emitted was pyroclasts with smaller amounts of lava flows than previously, forming a stratovolcano-type unit (Barrera Morate and García Moral 2011). The Ventejís-Pico-Moles group of volcanoes is composed of basaltic and tephritic lava flows, as well as basaltic tephra edifices with some hydromagmatic intercalations. These materials outcrop in the western and southeastern part of the town of Valverde (Gómez Sáinz de Aja et al. 2010) or as infill of between Tiñor and Tamaduste. The pyroclasts represented by bombs, slags, ashes and lapilli are olivine and tephritic basalts. Among the products from the late stage of the Tiñor formation, well-preserved volcanic cones with evidence of hydromagmatic interactions can be

distinguished (Barrera Morate and García Moral 2011). Finally, this stage ends with the Tiñor megathrust 0.88 My ago.

The rapid and voluminous growth of such a small area of the island caused the destabilization of the flank that originated the first gravity slide of the island in the NW of the newly formed edifice.

The later formative processes of the island have covered a large part of the Tiñor building although, thanks to the landslides, some of the materials of the three preceding stages have been left uncovered.

At present, the maximum height of this geological unit is 1137 m asl. There are materials from each of the formation sections: early stage/lower section, middle/tabular section and Ventejís section (Gómez Sáinz de Aja et al. 2010).

1.2 The Building El Golfo—Las Playas (0.54–0.18 My)

It is considered that at this time the volcanism associated with this building took place after the partial collapse of the

Tiñor Building, creating paleo-reliefs. Volcanic activity was mainly focused on the escarpment of the El Golfo arch to the north of the island and the midlands of the Las Playas arch to the southeast.

The outcrops of the units of this volcanic edifice are scarce as they have been covered by volcanic material from later events. The first section outcrops in the northern part of the island, in the Hoya del Verodal area and at the base of the Las Puntas cliff. As for the second section, it can be seen on the slopes of both escarpments, in El Golfo and Las Playas, resting discordantly on the materials of the lower section. In this way, it can be seen that this volcanism took place mainly on landslide escarpments. There is no clear evidence about the central area of this volcanic edifice, although with the study of the phyllonian network, there is the hypothesis that this area is close to the Cruz de los Reyes (Barrera Morate and García Moral 2011).

The orogenesis of this large volcanic edifice occurred during 0.35 My (0.54–0.18 My), in the Middle Pleistocene (Carracedo et al. 2001). The activity starts in the geological setting of the Tiñor edifice and has a NE-SW progression. All authors agree on the evolution of this volcanic edifice, although the age dating varies according to the area studied. After the volcanic inactivity of the Tiñor edifice, there was a reactivation that emitted basaltic and trachybasaltic lava flows and tephra cones discordant with the previous units. These types of emissions and units created are considered as the units of the lower section of the El Golfo—Las Playas edifice. Subsequently, the units of the upper section such as olivine-pyroxenic basaltic flows, olivine basalts, trachy-basalts, trachytic flows and mafic trachytes were deposited concordantly (Barrera Morate and Barrera Morate). mafic trachytes (Barrera Morate and García Moral 2011).

The lower section begins with the deposition of volcanic material on the slopes of the Tiñor landslide, which corresponds to the basaltic and hydromagmatic tephra cones. In addition, hydromagmatic volcanism occurred in some areas such as Sabinosa, leaving these buildings in discordance with the units above it.

The second section outcrops on the slopes of the escarpments of El Golfo and Las Playas. Both sections are dominated by basaltic, trachybasaltic and tephritic lava flows, the second section being differentiated by the presence of tephritic basanites and intercalations of basaltic, trachybasaltic, basanitic and tephritic tephra cones. In addition, in the upper part of the second section there are more acidic, alkaline and more evolved flows, such as trachytes and mafic trachytes (Carracedo et al. 2001).

The petrological composition is quite broad, although it depends on the study area and its emission centres. In general, there is an increase in alkalinity over time, with the last lava flows appearing to the SE of the Las Playas escarpment, dated at about 0.176 My (Barrera Morate and García Moral

2011). There is undoubtedly agreement between the different stages of formation of the island with some distinct substages depending on the author, although the petrological evolution varies a little more (Fuster et al. 1993; Carracedo et al. 1997; Guillou et al. 1996).

1.3 The Volcanism of the Rifts (0.15–0.012 My)

The volcanism of the rifts occurred 0.15 and 0.012 My ago along the three structural axes of the island (Fig. 2). It corresponds to the last stages of construction of El Hierro and affects much of its entire surface. The magma thrust broke the earth's crust creating a triple fracture (Carracedo et al. 2008). These fractures extend along the three main structural axes. These axes have a West-Northwest, North-Southwest and South-Southeast orientation, where they converge and are arranged in a series of eruptions that have shaped the Herreño territory up to the present day.

The activity of this last phase of formation began between 0.17 and 0.15 My ago. However, one of the most important processes related to this volcanism were the infill eruptions that occurred in the geographic setting of the El Golfo valley (Gómez Sáinz de Aja et al. 2010) and also in the afore-mentioned rift sectors, being a more moderate volcanism (Barrera Morate and García Moral 2011). The main characteristic of the volcanism in the interior of El Golfo is its location near the escarpment, as well as its duration, which is estimated to be no more than 10,000 years. The fact that there was little time for its formation is demonstrated by the fact that there are no important discordances or contrasts in its morphology (Barrera Morate and García Moral 2011). The eruptions that occurred during the Holocene of this phase led to the formation of lava deltas that gained ground to the sea (Barrera Morate and García Moral 2011) and caused the fossilization of many cliffs and the smoothing of the morphology of the coast.

In the rest of the Herreño territory where eruptions have occurred in this period, two stages can be distinguished: sub-recent and recent emissions (Barrera Morate and García Moral 2011). The former correspond to isolated eruptions in the structural axes of the three rifts. Some of the volcanoes belonging to this volcanism are the mountains Cueva del Guanche, Tomillar, Fara, Las Charquillas, Las Tabladas and Las Montañetas, among others (Barrera Morate and García Moral 2011). Recent emissions take place at the ends of the three rifts, in the western part (Punta del Verodal), in the northeast of the island (Hoya del Tamaduste) and in the south (La Restinga). All this activity meant, as mentioned above, an increase in the size of the island's perimeter due to the formation of the coastal platforms. The volcanoes that are part of these emissions are the Orchilla-Calderetón, Las Calcosas, Hoya del Verodal, Punta de la Dehesa,

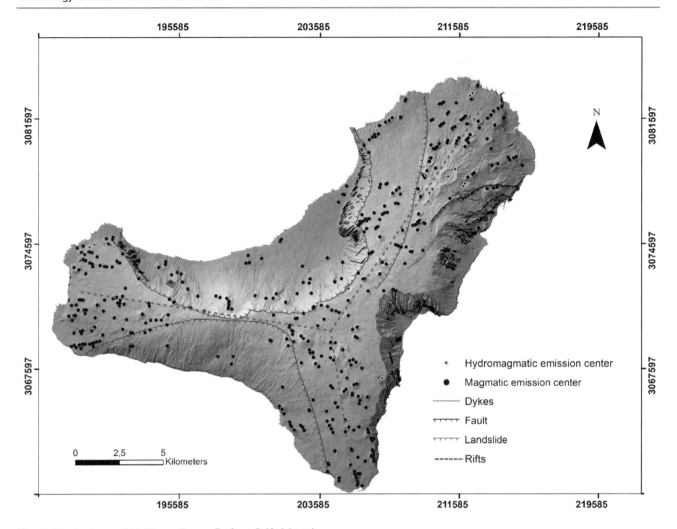

Fig. 2 Structuralmap of El Hierro. *Source* Grafcan. Self-elabortaion

Chamuscada-Entremontañas, La Cancela and Aguajiro mountains, among many others (Barrera Morate and García Moral 2011).

During the historical volcanic activity there is only one eruption in 2011 and a supposed eruption in the year 1793, although not considered so far as a historical eruption (Romero Ruiz 1989). This volcanic eruption that may have taken place in the historical past, has been dated with the carbon 14 method, which establishes it in the year 1793—it is the so-called "Lomo Negro" eruption (Barrera Morate and García Moral 2011).

The only historical eruption on the island began on October 10, 2011 and lasted until March 2012. It was a submarine eruption that originated 400 m deep and 1.5 km SSW of the coast of La Restinga. The originated volcano was named "Tagoro". Such volcanic eruption is the first of the twenty-first century in the Canary Islands, therefore, the study of the volcanic characteristics and its evolution are well documented (Barrera Morate and García Moral 2011; Pérez-Torrado et al. 2012; Domínguez Cerdeña et al. 2018;

Melián et al. 2014; Pérez et al. 2012, 2014, 2015; Padrón et al. 2013; Carracedo et al. 2012; Rivera et al. 2013). Particularly interesting were the surface seawater discoloration phenomena and the presence of floating volcanic material. These are dark crusted pyroclasts, formed by basanites, of basaltic character and white cores, rich in silica and very porous (Rodríguez-Losada et al. 2014; Pérez-Torrado et al. 2012) of trachytic-riolitic character, finally named as "Restingolites".

2 Conclusions

The island of El Hierro has gone through several formative stages over the last hundreds of thousands of years. These stages shaped the current relief, being interspersed between phases of formation and erosion to large dimensions. The first stage of the formation is known as the Tiñor volcanic edifice, composed of 3 formative sub-stages and a gravitational landslide that culminated its construction process.

After a long period of inactivity, the island began a new stage of growth (El Golfo). The new volcanic materials were deposited in discordance with the previous materials on the escarpments of El Golfo and Las Playas.

Finally, the last stage of the island's formation encompasses rift volcanism. This is a process characterized by two types of events, on the one hand the volcanic activity itself produced in the three structural axes of the island, as well as in the valley of El Golfo, and on the other hand, three gravitational landslides (El Julan, Las Playas and El Golfo) that determined the current configuration of El Hierro.

As the volcanic activity in the Canary Islands is understood, being El Hierro one of the western islands, the youngest and adding the last eruption in La Restinga, it is confirmed that the process of growth of the island has not yet finished.

References

Anguita F, Hernán F (2000) The Canary Islands origin: a unifying model. J Volcanol Geoth Res 103:1–26

Barrera Morate JL, García Moral R (2011) Geological map of the Canary Islands. General report. Coordinator Ricardo García Moral. Grafcan Ediciones

Carracedo JC, Day S, Guillou H, Badiola H, Rodríguez E, Canas A, Torrado J, Pérez L, Francisco M (1997) Geochronological, structural and morphological constraints in the genesis and evolution of the Canary Islands. In: International workshop, Sept 15–18: programme and abstracts, pp 45–48

Carracedo JC, Badiola ER, Guillou H, De la Nuez J, Pérez Torrado FJ (2001) Geology and volcanology of La Palma and El Hierro, Western Canaries. Estud Geol 57(5–6):175–273. https://doi.org/10.3989/egeol.01575-6134

Carracedo JC, Torrado FP, Badiola ER, Paris R (2008) Reliability of interpretation of subhistoric eruption references from Tenerife: the pre-Holocene eruption of Montaña Taoro. Geo-Temas 10. ISSN: 1567-5172

Carracedo JC, Torrado FP, González AR, Soler V, Turiel JLF, Troll VR, Wiesmaier S (2012) The 2011 submarine volcanic eruption in El Hierro (Canary Islands)

Day S, Carracedo JC, Guillou H (1997) Age and geometry of an aborted rift flank collapse: the San Andrés fault system, El Hierro, Canary Islands. Geolog Mag 134(4):523–537. https://doi.org/10.1017/S0016756897007243

Domínguez Cerdeña L, García-Cañada MA, Benito-Saz C, del Fresno H, Lamolda J, Pereda de Pablo C, Sánchez S (2018) On the relation between ground surface deformation and seismicity during the 2012-2014 successive magmatic intrusions at El Hierro Island, Tectonophysics, vol 744, pp 422–437. ISSN 0040-1951. https://doi.org/10.1016/j.tecto.2018.07.019

Fuster JM, Hernán F, Cendrero A, Coello J, Cantagel JM, Ancochea E, Ibarrola E (1993) "Geomorphology of El Hierro Island". Boletín de la Real Sociedad Española de Historia. Geol Sect 88:1–4

Gee MJR, Watts AB, Masson DG, Mitchell NC (2001) Landslides and the evolution of El Hierro in the Canary Islands. Mar Geol 177(3/4):271–293

Gómez Sáinz de Aja JA, Klein E, Ruiz García MT, Balcells Herrera R, Del Pozo M, Galindo E, La Moneda E (2010) Geological and Mining Institute of Spain National geological map (MAGNA), 1st edn, 2nd series

Guillou H, Carracedo JC, Pérez-Torrado FJ, Rodríguez-Badiola MI (1996) K–Ar ages and magnetic stratigraphy of a hotspot-induced fast-growing oceanic island: El Hierro, Canary Islands. J Volc Geoterma Res 73:141–155

Melián G, Hernández P, Padrón E, Pérez NM, Barrancos J, Padilla G, Dionis S, Rodríguez F, Calvo D, Nolasco D (2014) Spatial and temporal variations of diffuse CO_2 degassing at El Hierro volcanic system: relation to the 2011–2012 submarine eruption. J Geophys Res Solid Earth 119:6976–6991. https://doi.org/10.1002/2014JB011013

Padrón E, Pérez NM, Hernández P, Sumino H, Melián G, Barrancos J, Nagao K (2013) Diffusive helium emissions as a precursory sign of volcanic unrest. Geology 41(5):539–542. https://doi.org/10.1130/G34027

Pérez NM (2015) The 2011–2012 El Hierro submarine volcanic activity: a challenge of geochemical, thermal and acoustic imaging for volcano monitoring. Surtsey Res 13:55–69

Pérez NM, Padilla G, Padrón E, Hernández PA, Melián G, Barrancos J, Samara D, Nolasco D, Rodríguez F, Calvo D, Hernández I (2012) Precursory diffuse CO_2 and H_2S emission signatures of the 2011–2012 El Hierro submarine eruption, Canary Islands. Geophys Res Lett 39(16) ISSN:0094-8276. https://doi.org/10.1029/2012GL052410

Pérez NM, Somoza L, Hernández PA, González de Vallejo L, León R, Sagiya T, Biain A, González FJ, Medialdea T, Barrancos J, Ibáñez J, Sumino H, Nogami K, Romero C (2015) Reply to comment from Blanco et al. on "Evidence from acoustic imaging for submarine volcanic activity in 2012 off the west coast of El Hierro (Canary Islands, Spain) by Pérez et al. [Bull Volcanol 76:882–896]". Bull Volcanol 77(7):1–8. https://doi.org/10.1007/s00445-015-0948-5

Pérez NM, Somoza L, Hernández PA, González de Vallejo L, León R, Sagiya T, Biain A, González FJ, Medialdea T, Barrancos J, Ibáñez J, Sumino H, Nogami K, Romero C (2014) Evidence from acoustic imaging for submarine volcanic activity in 2012 off the west coast of El Hierro (Canary Islands, Spain). Bull Volcanol 76:882. https://doi.org/10.1007/s00445-014-0882-y

Pérez-Torrado FJ, Carracedo JC, Rodríguez-González A, Soler V, Troll VR, Wiesmaier S (2012) The submarine eruption of La Restinga on the island of El Hierro, Canary Islands: October 2011–March 2012. Geol Stud 68(1):5–27. https://doi.org/10.3989/egeol.40918.179

Rivera J, Lastras G, Canals M, Acosta J, Arrese B, Hermida N, Micallef A, Tello O, Amblas D (2013) Construction of an oceanic island: Insights from the El Hierro (Canary Islands) 2011–2012 submarine volcanic eruption. Geology 41(3):355–358. https://doi.org/10.1130/G33863.1

Rodriguez-Losada JA, Eff-Darwich A, Hernandez LE, Viñas R, Pérez NM, Hernandez P, Melián G, Martinez-Frías J, Carmen Romero-Ruiz M, Coello-Bravo JJ (2014) Petrological and geochemical Highlights in the floating fragments of the October 2011 submarine eruption offshore El Hierro (Canary Islands): relevance of submarine hydrothermal processes. Afr Earth Sci. https://doi.org/10.1016/j.jafrearsci.2014.11.005

Romero Ruiz C (1989) Las manifestaciones volcánicas históricas del archipiélago canario. University of La Laguna. ISBN: 84-7756-219-9

Sandoval-Velasquez A, Rizzo A, Aiuppa A, Remigi S, Padrón E, Pérez NM, Frezzotti ML (2021) Recycled crystal charcoal in the depleted mantle source of El Hierro volcano, Canary Islands. Lithos. https://doi.org/10.1016/j.lithos.2021.106414

Volcanic Geomorphology in El Hierro Global Geopark

Cayetano Guillén-Martín and Carmen Romero

Abstract

Few oceanic islands express their geomorphological history in such a marked way as the island of El Hierro. Indeed, on El Hierro, its geomorphology goes hand in hand with the evolution of its insular geology. In fact, seventy percent of places of geological interest in El Hierro's Geopark have geomorphological features as their main or secondary interest, which is indicative of the importance of geomorphology in the configuration of the island's relief. However, there are few studies that have addressed the processes or features of the island's geomorphology. In this study, the first geomorphological characterization is carried out in which the island is considered as a whole unit.

Keywords

Volcanic geomorphology · Monogenetic volcanism · Geopark · El Hierro · Canary Islands

1 Introduction

With a maximum age of 1.2 Ma (Guillou et al. 1996), El Hierro is the youngest island of the Canary Archipelago. It is an oceanic volcanic island formed by the fusion of the Tiñor

C. Guillén-Martín (✉)
EUTUR, University School of Tourism, Santa Cruz de Tenerife, Spain
e-mail: cayetano.guillen@eutur.es

C. Romero
Department of Geography and History, University of La Laguna, San Cristóbal de La Laguna, Spain

C. Guillén-Martín · C. Romero (✉)
GPS-VOLTER Geomorphology, Landscape and Society in Volcanic Territories, University of La Laguna, San Cristóbal de La Laguna, Spain
e-mail: mcromero@ull.edu.es

(1.12–0.88 Ma) and El Golfo-Las Playas (545–176 ka) volcanic edifices, as well as by the Holocene volcanic fields that developed later. Several structures can be identified in the island's morphology that have been interpreted as the scars of giant gravity landslides of Tiñor (0.8–0.5 Ma), Las Playas (~ 545–176 ka and 176–145 ka), El Julán (> 158 ka), and El Golfo (~ 87–39 ka) (Carracedo et al. 1999, 2001; Gee et al. 2001; Longpré et al. 2011; Masson 1996; Masson et al. 2002). The youngest of these landslides corresponds to a broad amphitheatre open to the NW and bounded by a large 27 km long arcuate escarpment that gives the island its distinctive crescent shape. From about 158 ka ago, monogenetic volcanism has developed on the flanks of the previous structures and in the interior of the depressions generated by landslides, and whose distribution is controlled by an apparent triaxial system of volcanic rifts (Carracedo et al. 2001; Guillou et al. 1996).

From a climatic point of view, the island of El Hierro has an oceanic subtropical climate, with temperatures ranging between 19 and 23 °C and rainfall concentrated from October to March. In addition, rainfall varies by slope depending on the slope's exposure to the prevailing winds, though it can exceed 1000 mm per year in windward areas.

The climate of El Hierro results from the interaction between the general climatic conditions of the whole archipelago and the island's steep mountainous relief. The dominant trade winds on the island reach El Hierro via the eastern flank of the Azores anticyclone of moderate speed (20–25 km/h) (Marzol 2006), which brings humidity and are present in the archipelago for almost two thirds of the year.

The combination of these elements gives rise to a relatively complex mosaic of microclimates, with different predominant morphodynamic processes, ranging from those typical of temperate-humid climates (areas between 800 and 1500 m asl. on the N and NE slopes of the island) to those associated with semi-arid climatic contexts (coasts and slopes of the S, SW and W of El Hierro).

© The Author(s) 2023
J. Dóniz-Páez and N. M. Pérez (eds.), *El Hierro Island Global Geopark*,
Geoheritage, Geoparks and Geotourism, https://doi.org/10.1007/978-3-031-07289-5_3

The cartography that accompanies this study was conducted taking into consideration the fieldwork carried out on the island in recent years, a high resolution DEM analysis, as well as the interpretation of the aerial photographs of the island from recent decades, available through the IdeCAN WMS server (https://www.idecanarias.es/). The different geographical features were mapped and classified according to their volcanic, erosive or sedimentary genesis.

The delineation of the riverbeds has been carried out manually in a GIS environment following the method of Strahler (1964), since important areas of the island characterized by lava surfaces are structured in slight and central channels and areas of interlavic and intralavic contact gave very high errors in their automated digital delineation. Nevertheless, the delimitation of the proposed mapping of the basins is based on the information derived from slope models, high resolution DEM and the runoff accumulation map.

The study of erosive forms has been based on the morphometric analysis of the hydrographic network. This not only provides a quantitative description of the drainage system, but also constitutes one of the tools that gives most information for the morphological study of volcanic territories. Basins and talwegs effectively express the existing relationships between the lithological, structural, geomorphological and climatic characteristics of the territory (Romero et al. 2004, 2006).

The study of volcanic forms has been carried out through the mapping of volcanic vents and cones and the morphometric analysis of some of their most relevant parameters, such as the elongation of craters and direction of volcanic fissures (Dóniz 2004, 2008). For the morphological classification of the cones, the taxonomy of Bishop (2009) has been followed. In his morphometric and geomorphological study, surtseian cones (Romero 2016; Guillén, in press) or those highly disfigured by erosion or landslides have not been included.

2 Physiographic Features

The island of El Hierro has an abrupt and vigorous relief of marked orographic contrasts. Twenty percent of the island's surface is characterized by an open depression to the NNW, closed to the south by a steep escarpment with slopes of up to 600 m in height, and where the highest altitude of the island is reached (1501 m at the Pico de Malpaso). This great amphitheatre is one of the island's most characteristic features. From this cliff, the altitude of the island descends quite steeply in all directions, forming ramps that extend fundamentally towards the interior of the depression, and towards the NE, S and W (Fig. 1a).

However, altitudes, slope breaks and slope distribution allow us to divide the island into eight different areas (Fig. 1b). Each of these areas has specific physiographic features and represent, in relation to their orientation towards the trade winds, and their more or less marked altitude, specific bioclimatic and morphoclimatic units. The north-facing units of Valverde (VA), Tiñor (TI), Nisdafe (ND) and El Golfo (EG) (Fig. 1b, nos. 1, 2, 3 and 5) are open slopes to the NE and N, more humid, with higher rainfall and lower insolation, where weathering processes predominate. The leeward areas of Las Playas (LP), El Pinar (EP), El Julan (EJ) and La Dehesa (LD) (Fig. 1b, nos. 4, 6, 7 and 8) are characterised by a warmer climate, with less rainfall and cloud cover and fewer mechanical erosion processes.

3 Erosive and Accumulation Forms

Even though the ravines do not seem to be one of the outstanding features of the relief of El Hierro, given the youth of the island, the island has a well-established hydrographic network, consisting of 270 basins and 3683 channels, with an average density of 5 km/km^2. The low rainfall, high evaporation rates and the predominance of very permeable materials, determines that there is no water regime with permanent flows on the island. These are intermittent and ephemeral watercourses, usually with sporadic torrential flows, associated with the development of high intensity rainfall episodes (Marzol 2006; Arroyo 2009). Figures 1 and 2 summarize the most salient features of drainage on an island scale.

The orographic and geological setting of the riverbeds and basins means that 72.4% of the island's basins are radially distributed with respect to the EG escarpment and are cataclysmic, running down the dip of the layers. Forty-one percent of the basins have their headwaters on the EG ridgeline, which acts as the main watershed of the island. This watershed extends to the NE following the ridge line established around the Ventejís volcanic vent (1238 m) (Fig. 3a–c).

As on other islands of the Canary archipelago (Romero et al. 2006), the control imposed by the lithology in combination with the age of the materials conditions the existence of a very significant surface without drainage network, which extends 78 km^2. In addition, at least 31% of the island's basins have no direct outlet to the sea and can be considered endorheic basins. Although surface without drainage network and endorheism affect the whole island, they are fundamental features of the eastern area of the EG amphitheatre and the volcanic fields of EP and LD (Fig. 2d). The presence of these relict and endorheic spaces is not directly linked to environmental factors but is determined by the inhibition of runoff in areas of more recent volcanism, caused by the high porosity and permeability of the volcanic materials (Romero et al. 2006).

The number and shape, order and area of the river basins show spatial variations and define very diverse hydrographic

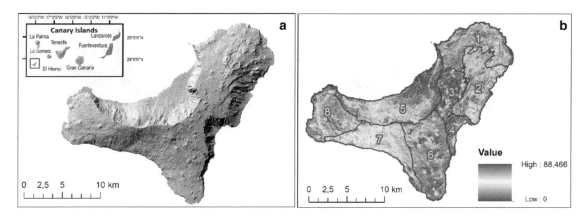

Fig. 1 Digital shadow model of the island of El Hierro (**a**) and slope map (**b**), with the delimitation of the physiographic units, 1: Valverde (VA); 2: Tiñor (TI); 3: Nisdafe (ND); 4: Las Playas (LP); 5: El Golfo (EG); 6: El Pinar (EP); 7: El Julan (EJ) and 8: La Dehesa (LD)

	km2	Stream Order 1	Stream Order 2	Stream Order 3	Stream Orderr 4	Stream Order 5	Talwegs Length km	Density km /km²
Island	268	2659	745	184	45	5	1348,3	5,0
Valverde	45	198	44	10	3	-	134,0	3,0
Tiñor	42	427	129	28	6	1	225,0	5,4
El Golfo	55	477	143	38	6	-	184,4	3,4
Las Playas	19	466	117	31	6	1	165,7	8,7
El Pinar	41	248	71	12	3	-	128,1	3,1
El Julan	42	680	193	53	17	3	394,7	9,4
La Dehesa	25	163	48	12	4	-	116,5	4,7

	Basins	Basin order 1	Basin order 2	Basin order 3	Basin order 4	Basin order 5	Endorheic basin
Island	270	76	98	60	32	5	85
Valverde	19	8	6	2	3	-	1
Tiñor	37	10	11	11	4	1	6
El Golfo	74	18	31	19	6	-	55
Las Playas	31	4	12	11	3	1	-
El Pinar	26	8	12	3	3	-	13
El Julan	63	20	20	11	9	3	2
La Dehesa	20	7	6	3	4	-	8

Fig.2 Data relating to the hydrographic network and watersheds of the island of El Hierro

units that adapt to the different geological units, coinciding roughly with the physiographic units mentioned above. Although topographically ND is an area with its own characteristics, hydrographically it is part of the TI and VA areas, as the headwaters of their watercourses are located in this sector.

The island's watersheds show elongated shapes and lack, except in the steep sectors of EG and LP and TI, reception areas with clear topographic limits. The highest density of watercourses corresponds to EJ and LP, with values of 9.4 and 8.7 km/km², respectively. The lowest values characterize the VA, EP, LD and EG sectors (with 3, 3.1, 4.7 and 3.4 km/km² respectively). TI has average drainage densities of 5.4 km/km².

In general, lithostratigraphic variations, discontinuities in the rocks (joints and internal structure of the lava flows), dips of the layers in relation to the direction of runoff, tectonic features, age of the materials, morphoclimatic environment, and degree of interference between volcanic and erosive processes act in an interrelated way to give rise to valleys and ravines of very diverse morphology.

In the areas corresponding to the old massif of TI or in the sliding escarpments of EG and LP, where the oldest outcrops of basalts are on the island, the most important levels of network encasement are reached. In the TI area, where interference with later volcanism has been practically nil, erosion has dismantled the original structures and has generated the presence of abrupt reliefs, with steep slopes and deep torrential incisions separated by interfluves on ridges. These are inverted reliefs, which characterize the middle and lower sections of the basins, carved at the expense of the piles of the lava flows of the lower sequence of the IT edifice. The upper sections of these basins show, however, significantly lower levels of wedging when adapting to the tabular piles of the intermediate sequence. The transition between the two geological units is marked by the presence

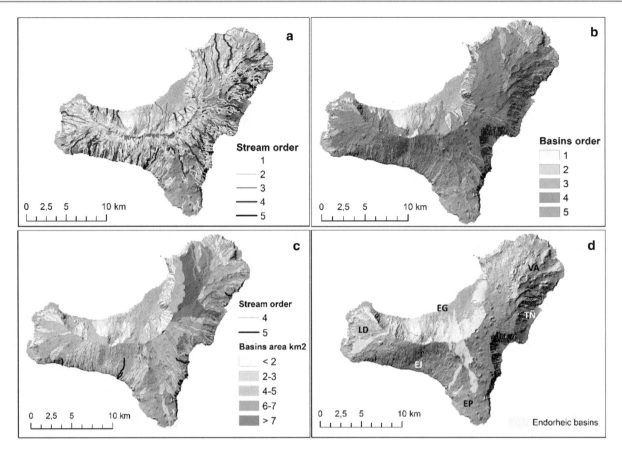

Fig. 3 Distribution of watercourses (**a**) and basins (**b**) according to their order. Classification of the basins according to their area in km² (**c**). The areas without colour in maps **a–c** correspond to reef sectors, without an organized drainage network. Map **c** shows the basins that do not flow into the sea and can be considered endorheic

of abrupt jumps and pronounced slope breaks in the profile of the main valleys. Between the main interfluves of these basins, parallel ridge interfluves develop, with summit-to-slope gradients of 200–300 m difference between the summit and interfluves. To the south of TI, later lava flows completely flooded the channels and partially clogged the basins, leaving the ridge interfluves as isolated remnants, causing the appearance of channels with a lesser degree of embedding in contact areas with other geological units. This determines that TI is the only area of the island where differentiated incision levels and the development of quadrangular-shaped basins can be defined, with a drainage pattern with dendritic trends and morphological evidence of recent landslide (Klimeš et al. 2016), with pulses linked to climatic events (Blahut et al. 2018).

The combination of high slopes and the structure of the EG and LP landslide escarpments has favoured very effective erosion action. The configuration of the escarpments in layers with different mechanical properties (Isidro et al. 2015), with numerous discontinuity planes (alternation of lava and pyroclastic layers, joints and internal levels of the lava layers, degree of weathering and presence of subvertical

dykes) has favoured the presence of rocky channels vertically embedded in the walls, locally called "*fugas*". These channels are characterised by high hypsometric gradients, short but very pronounced longitudinal profiles and frequent cornices and breaks in the slope. They show funnel morphologies that favour the channelling of landslides and the accumulation of sediments at the foot of the most vertical escarpments.

The difference in age and dip of the layers in relation to the direction of the runoff, fundamentally anaclinal in EG and cataclinal in LP, determine variable degrees of remodelling between the escarpments of both depressions. These variations are evident in the different degrees of channel embedding, the topographic definition of the interfluves and the catchment basins and the lobulated character of the general outline of the escarpments, more accentuated in the case of LP than in EG. The degree of erosive remodelling of these escarpments has been established by calculating their sinuosity index, which is the result of dividing the total length of the escarpment by the straight line length of its starting and end points. Thus, while the LP escarpment has a sinuosity index of 2.38, in EG it is only 0.72. These data are indicative of the greater degree of erosive evolution of the LP escarpment compared to that of EG, which fits with the chronology of both depressions. The ruiniform character of the relief of the LP amphitheatre is also associated with the development of an anaclinal hydrographic network with its catchment areas on the upper slopes and the absence of subsequent monogenetic volcanism in the interior of the depression.

The gullies developed in the areas of monogenetic volcanism (volcanic fields of the NE, S, W) or in the EJ landslide arc are usually characterized by their low degree of embedding, and by generating elongated and narrow basins, usually with slightly lobulated headwaters and with surfaces of less than 4 km^2. In these sectors, the pyroclasts that characterize the summit areas favour the formation of very dense networks in the headwater areas, with a multitude of small length channels of order 1 and 2. On the other hand, the steep slopes, essentially formed by lava flows structured in channels and lateral levees, favour the development of radial or parallel networks, with a tendency to be embedded in lava channels or in the contact areas between different lava units. These networks are not very hierarchical and have areas of less than 2 km^2, where long and narrow basins of orders 2 and 3 predominate.

The anomalies of the drainage network in the size and shape of the basins and in the layout of the network show a close relationship with tectonic processes, as well as the closing and blocking of the valleys associated with the emplacement of cones and lava flows. The role of tectonics can be observed especially in two points of the island: in LD, in the transit zone towards EJ and in some areas of TI. The influence of tectonics is reflected in the existence of breaks in the profile of the channels and in the design of their layout. In these sectors, there are frequent abrupt changes in the course of the channels, with zigzag or staircase traces and steps in the profile that allow the different fault breaks to be bridged. These features can also be observed in the network around the Caldera de Ventejís, where the talwegs with the most severe boxed are associated with the presence of faults.

The processes of closure and obturation of valleys and basins cause processes of confluence and difluence of the ravines that determine significant changes in the layout of the network and in the size of the basins, favouring the presence of basins with surfaces greater than 6 km^2 and the existence of plains and endorheic type areas.

The products of erosion are preserved as spatially discontinuous sedimentary units, characterizing the mouth areas of the steepest ravines, or those with abrupt breaks in slope in their lower sections, where a range of detrital fans are formed (Fernández-Pello 1989). Many of these deposits, however, lack the typical morphology of fans as they adapt to the irregular surface of recent lava flows, where they generate the appearance of irregular fluviotorrential fans. Also, at the foot of the large escarpments of the LP and EG

depressions, or at the bottom of the larger cliffs surrounding the island, there are detrital fans of mixed origin, the result of both gravity processes and small debris-flow processes.

In the deposits at the foot of the slope of the LP and EG escarpments, it is possible to distinguish two or three generations, with topographic locations that are lower in altitude, the more recent they are. In general, the oldest deposits are relicts and appear to be dissected by boxed channels, which have ended up disconnecting the apexes and bodies of the fans from the current escarpments, while being topographically disconnected from their source areas. When they are located on the coast, their bases are cliffed by the sea, so their formation must have taken place at a lower sea level than the present one. These deposits are flanked and covered at their base by the most recent deposits of the second or third generation, so they have more gradual average slopes.

4 Monogenetic Mafic Volcanism

The monogenetic basaltic volcanism of El Hierro characterizes the last stage of construction of the island, concentrating along three zones identified as volcanic rifts (Carracedo 1994; Balcells et al. 1997, Aulinas et al. 2019). The 541 mapped emission centres depict an essentially radial pattern with respect to the EG arc (Becerril 2014; Becerril et al. 2015, 2016) with an average density of 2.02 vents km^2. The ND, EP and LD areas (areas 3.6 and 8 in Fig. 1) show values above the island mean, with 3.7; 3 and 5.6 vents/km^2 respectively. By contrast, some areas, such as TI and EJ, have very low vent densities, with 0.54 and 0.49, respectively (Fig. 4).

Each of these areas presents a predominant orientation around one or two preferred directions that help to delimit the volcanic fields on the island. Thus, the volcanic field of LD can be structurally divided into two sectors according to the number of emission centres, their spatial distribution and the greater or lesser predominance of specific directional lines. On the one hand, the WNW sector has a significantly lower number of vents (54 compared to 115 in the WSW sector). There are, however, many very marked eruptive fissures aligned following the predominant NW–SE and WNW-ESE directions. This sector forms a group of volcanic cones spatially separated from the rest of the cones in the western sector of the island. The WSW sector, on the contrary, shows not only a greater number of eruptive centres, but also its organization is via the preferential fractures of ENE, WSS. However, NW–SE and WNW-ESE orientations are also present in a less representative way.

Something similar occurs in the NE part of the island, where the distribution and geometry of the cones and fissures define at least two sectors with different features. On the one hand, the areas of VA and TI (Fig. 1, zones 1 and 2),

Morphological units	Km²	Nº Vents	Density vents/km²	Nº Cones	Density cones/km	Nº vents/cones	Simples	% with respect to total cones	Compounds	% with respect to total cones	Complex	% with respect to total cones
TOTAL ISLAND	268	541	2,02	230	0,86	2,35	68	34,87	99	50,77	28	14,3589744
Valverde	37	66	1,78	36	0,97	1,83	13	6,67	13	6,67	5	2,56
Tiñor	37	20	0,54	12	0,32	1,67	2	1,03	3	1,54	4	2,05
Nisdafe	18	68	3,78	26	1,44	2,62	7	3,59	13	6,67	2	1,03
El Golfo	54	54	1,00	25	0,46	2,16	5	2,56	12	6,15	5	2,56
Las Playas	9	-	-	-	-	-	-		-		-	
El Pinar	48	147	3,06	65	1,35	2,26	21	10,77	28	14,36	7	3,59
El Julan	35	17	0,49	12	0,34	1,42	-		-		-	
La Dehesa N	11	54	4,91	14	1,27	3,86	5	2,56	9	4,62		0,00
La Dehesa S	20	115	5,75	41	2,05	2,80	15	7,69	21	10,77	5	2,56

Fig. 4 Parameters for the characterization of monogenetic edifices

where the cones seem to be arranged following a marked NE-SW direction. On the other hand, the area of ND, where the cones are articulated in an arched shape with respect to the EG escarpment. This arcuate configuration is also evident around the EJ arc, the preferential orientations seem to be arranged in an arc following its east and west limits. This affects the WSW and S volcanic fields, which lose their radial distribution and do not configure well-defined volcanic rifts structurally.

A total of 230 monogenetic volcanic edifices have been mapped, in different locations. They have a range of ages, morphologies, sizes, materials and geometries. All of them are associated with mafic basaltic magmas, show accentuated fissural features (with an average of 2.3 vents per volcanic cone) and have eruptive mechanisms that vary from typically Hawaiian and Strombolian to violent Strombolian, and eventually with the development of phreatomagmatic pulses. They correspond to spatter, slag and lapilli cones that emitted abundant lava flows.

Volcanoes have been grouped, following the taxonomy established by Bishop (2009), into three categories: *simple, compound and complex*. Some 34.8% are simple cones, characterized by being unique and discrete edifices, spatially isolated and without interaction with other cones. Their morphology can be annular or horseshoe-shaped, with a simple ground plan. There are 54.7% that correspond to compound cones in which two or more cones of the same type are merged, so they usually show coalescent or multiple craters (Tibaldi 1995; Corazzato and Tibaldi 2006) and lobulated and irregular plans. Finally, 14.3% are complex edifices, resulting from the combination or superposition of two or more types of volcanoes with differentiated eruptive mechanisms. This last category includes mainly those cones built by slag, lapilli or spatter volcanic edifices associated with basal effusive vents of composite pahoehoe flows. These effusive emission centres have given rise to small lava shields (scutulum type) (Walker GPL, 2000), some of which seem to correspond to rootless secondary lava edifices built on volcanic tubes. An outstanding feature of these small lava shields is their high average slope, as the lavas are usually in areas with steep pre-slopes. These features are particularly characteristic of the monogenetic edifices of Holocene platform-forming volcanism, but they can correspond to volcanic assemblages of very different ages. These features can be found in VA (Montaña del Tesoro), in TI (such as in WSW (Montaña de Orchilla, Montaña de Lomo Negro, Montaña de Las Calcosas, Montaña Tenaca, Montaña Quemada de Allá), or in the S rift (Montaña de El Julan, Montaña de La Empalizada, etc.) and even within the EG depression (Montaña de Sabinosa, Tanganasoga). The volumes of these lava edifices are generally higher than those of the cones from which they originate.

The link between the cones and their corresponding lava flows is not always visible. The intensity of the degradation processes determines that the connection between the older flows, and the edifice is difficult to establish, as it is diffuse in morphoclimatic environments of high humidity, where weathering processes predominate over erosive ones, and where there is also a high degree of anthropization of the territory (Fig. 5).

Many of these flows are partially shaped by erosion and sedimentary processes, so that it is possible to group them into at least four categories. The first category consists of lava flows with a very low degree of preservation and whose surface morphology has been lost and cannot be mapped. However, in this category, we have also differentiated from those which, although showing very low degrees of preservation, can be mapped thanks to the existence of coastal platforms and the erosion carried out in the areas of contact with older materials. Next, there are lava flows with a medium degree of preservation still maintain well-defined morphological elements, such as slight laterals and fronts, pressure arches, etc., though their surface shows evident signs of erosion. They are located at the bottoms of ancient channels, as well as in the contact zones between different lava flows. The surfaces of these lava flows are often covered with detritic materials that blur their original surface morphology, showing highly variable degrees of transformation. In the third category, there are flows with a high degree of preservation belonging to Holocene eruptions on

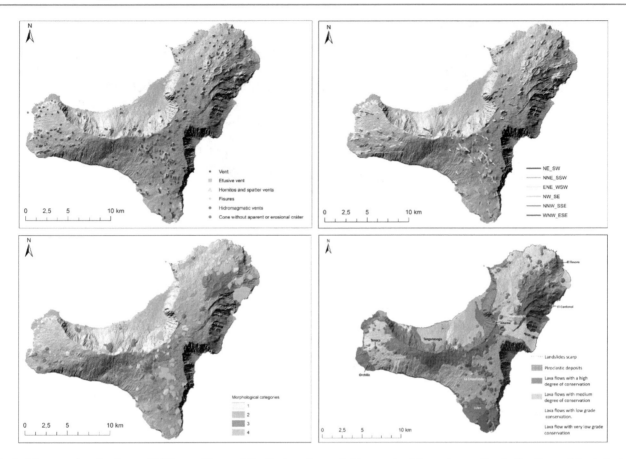

Fig. 5 Monogenetic volcanism on El Hierro. **a** Type and distribution of vents. **b** Lineaments of the vents according to directions. **c** Morphological types of volcanic cones. **d** Degree of preservation of the main lava units on the island

the coast of the northern humid and southern dry climatic environments, typical of the lava platforms and deltas of these areas. Their location at the foot of pre-coastal escarpments, however, favours an intense partial covering with detrital materials that are transported from talwegs established on the upper slopes and spill down to reach the coast. Finally, the best-preserved lava fields are located in the western and southern areas of the island, where xeric climatic conditions have led to the best preservation of the original forms.

5 Discussion and Conclusions

The delimitation of geomorphological units is one of the key factors in the geodiversity of a territory (Serrano and Ruiz-Flaño 2007). This study establishes at least eight geomorphological units on the island of El Hierro that result from the combination of topographic, structural, geological and geomorphological elements and factors. The temporal and spatial relationships between the processes of volcanic

construction and those linked to their erosive dismantling, or the formation of detrital deposits are also relevant.

Traditionally, the essential geomorphological features of El Hierro have been described using a simple model based on the three volcanic rifts that concentrate and organize the volcanic activity of the island around a triaxial scheme. This approach leads to structures with relatively simple topographic and geomorphological features, in whose interior the emission centres are concentrated in the central axis of these structures, giving rise to volcanic lineaments and obvious crater elongations, coinciding in orientation with a well-defined main structural orientation. However, it has been pointed out that the geometry of the volcanic fields on the island of El Hierro shows radial patterns that are a reflection of the local stress fields related to the formation of megathrust landslides that mask the general radial patterns of the EG edifice (Fig. 6).

This work contributes to define the geometry of the volcanic fields of the island of El Hierro, understood as the result of complex patterns and temporal interactions between the relative location of the different volcanic edifices that

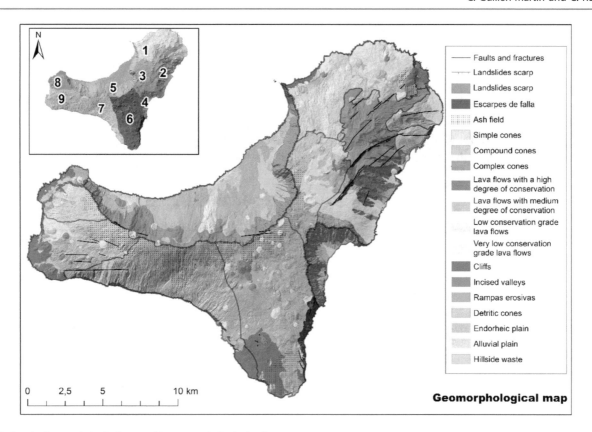

Fig. 6 Synthesis morphological map with geomorphological units

make up the island. Their morphological evolution and the regional geodynamics of the volcanic province are also relevant factors, as has been previously highlighted by authors for other Canary Islands (Márquez et al. 2018).

Structurally, the volcanic rifts of the Canary Islands have been defined as polygenetic edifices generated by the preferential emission of magma through persistent tectovolcanic fissures and associated with swarms of feeder dykes, whose density increases towards the rift axis and at depth (Carracedo 1994). Morphologically, these types of structures are characterized by the presence of narrow and longitudinally developed volcanic ridges orientated around a preferential directrix, with a well-defined line of summits located on its axis and with slopes built essentially by piles of lava flows with a generalized dip perpendicular to the axis. This gable roof configuration is palpable in the volcanic rifts of the islands of Tenerife and La Palma, although with nuances depending on the geological history of each edifice, its age and the greater or lesser degree of interaction between volcanic and erosive processes.

The variations between some volcanic fields and others and the comparative synthetic analysis of the morphological

features that characterize the volcanic rifts of Tenerife, La Palma and El Hierro show that only the NE volcanic field seems to correspond to a developed and persistent volcanic rift over time (Fig. 7, profile A–A′), although influenced by the stress field of the EG edifice in the ND zone (Fig. 7, profile B–B′). The WNW volcanic field corresponds to the volcanism controlled by the EG system directly on its flanks, without the construction of a well-defined axis. The S and WSW alignments have complex patterns that show the high influence of the stress field of the EG edifice and, above all, of the landslide effect. Some of these volcanoes are distributed by drawing arcs with respect to the EG landslide scarp or the EJ slopes, following approximately arcuate fractures towards the interior of both sectors. All these factors determine that the volcanic rifts do not have the morphological features in plan and elevation of other volcanic rifts of the Canary Islands, such as those of La Palma or Tenerife. This lack of morphological regularity of El Hierro's rifts cannot be linked to the age of the structures since the S rift of the island of La Palma has a similar age to the volcanism of the volcanic fields of El Hierro.

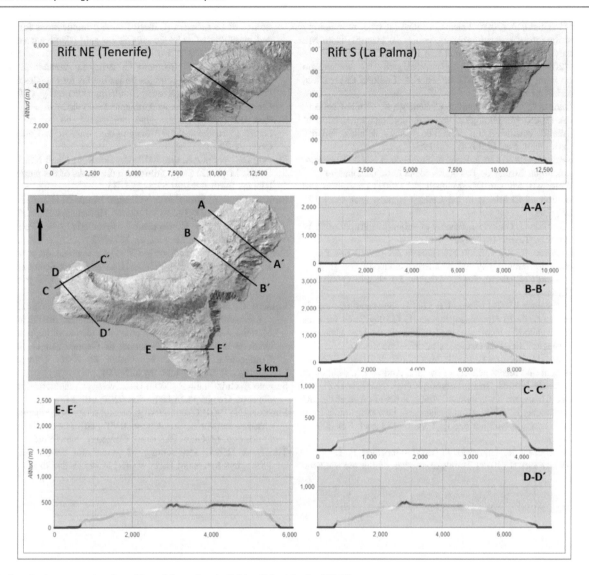

Fig. 7 Standard cross-sectional profiles of the volcanic fields of the island of El Hierro

References

Aulinas M, Domínguez D, Rodríguez-González A, Carmona H, Fernández-Turiel JL, Pérez-Torrado FJ, D'Antonio M (2019) The Holocene volcanism at El Hierro: insights from petrology and geochemistry. Geogaceta 65:35–38

Arroyo J (2009) Cinco siglos de la temperie canaria: cronología de efemérides meteorológicas. Asociación Canaria de Meteorología (ACANMET), Santa Cruz de Tenerife, Spain

Balcells R, Gómez JA (1997) Memorias y mapas geológicos del Plan MAGNA a escala 1: 25.000. El Hierro Island: Hoja 1105-III, Sabinosa. Spanish Geological Survey, Madrid, p 71

Becerril L (2014) Volcano-structural study and long-term volcanic hazard assessment on El Hierro Island (Canary Islands), Doctoral thesis. University of Zaragoza

Becerril L, Galindo I, Martí J, Gudmundsson A (2015) Three-armed rifts or masked radial pattern of eruptive fissures? The intriguing case of El Hierro volcano (Canary Islands). Tectonophysics 647:33–47

Becerril L, Ubide T, Sudo M, Martí J, Galindo I, Galé C, Lago M (2016) Geochronological constraints on the evolution of El Hierro (Canary Islands). J Afr Earth Sci 113:88–94

Bishop MA (2009) A generic classification for the morphological and spatial complexity of volcanic (and other) landforms. Geomorphology 111(1–2):104–109

Blahut J, Baroň I, Sokol' L, Meletlidis S, Klimeš J, Rowberry M, Marti X (2018) Large landslide stress states calculated during extreme climatic and tectonic events on El Hierro, Canary Islands. Landslides 15(9):1801–1814

Carracedo JC (1994) The Canary Islands: an example of structural control on the growth of large oceanic-island volcanoes. J Volcanol Geoth Res 60(3–4):225–241

Carracedo JC, Day SJ, Guillou H, Torrado FJP (1999) Giant quaternary landslides in the evolution of La Palma and El Hierro, Canary Islands. J Volcanol Geoth Res 94(1–4):169–190

Carracedo JC, Rodriguez-Badiola E, Guillou H, Nuez Pestana JDL, Pérez Torrado FJ (2001) Geology and volcanology of la Palma and El Hierro, western Canaries. Geol Stud 57:175–273

Corazzato C, Tibaldi A (2006) Fracture control on type, morphology and distribution of parasitic volcanic cones: an example from Mt. Etna, Italy. J Volcanol Geothermal Res 158(1–2):177–194

Dóniz J (2004) Caracterización geomorfológica del volcanismo basáltico monogénico de la isla de Tenerife (Doctoral dissertation, Universidad de La Laguna)

Dóniz J, Romero C, Coello E, Guillén C, Sánchez N, García-Cacho L, García A (2008) Morphological and statistical characterisation of recent mafic volcanism on Tenerife (Canary Islands, Spain). J Volcanol Geoth Res 173(3–4):185–195

Fernández-Pello L (1989) Los paisajes naturales de la isla de El Hierro. Excmo. Cabildo Insular de El Hierro. Santa Cruz de Tenerife

Gee MJ, Watts AB, Masson DG, Mitchell NC (2001) Landslides and the evolution of El Hierro in the Canary Islands. Mar Geol 177 (3–4):271–293

Guillen C, Romero C, Galindo I. Review of submarine eruptions in El Hierro prior to Tagoro (in Press)

Guillou H, Carracedo JC, Torrado FP, Badiola ER (1996) K–Ar ages and magnetic stratigraphy of a hotspot-induced, fast-growing oceanic island: El Hierro, Canary Islands. J Volcanol Geoth Res 73(1–2):141–155

Isidro ML, Jiménez IG, Bartolomé CP (2015) Desprendimientos de rocas en la Isla de El Hierro. In: Ingeniería Geológica en terrenos volcánicos: métodos, técnicas y experiencias en las Islas Canarias. Ilustre Colegio Oficial de Geólogos, pp 333–366

Klimeš J, Yepes J, Becerril L, Kusák M, Galindo I, Blahut J (2016) Development and recent activity of the San Andres landslide on el Hierro, Canary Islands, Spain. Geomorphology 261:119–131

Longpré MA, Chadwick JP, Wijbrans J, Iping R (2011) Age of the EG debris avalanche, El Hierro (Canary Islands): new constraints from laser and furnace 40Ar/39Ar dating. J Volcanol Geoth Res 203(1–2):76–80

Márquez A, Herrera R, Izquierdo T, Martín-González F, López I, Martín-Velázquez S (2018) The dyke swarms of the old volcanic Edifice of La Gomera (Canary Islands): implications for the origin and evolution of volcanic rifts in oceanic island volcanoes. Glob Planet Change 171:255–272

Marzol MV, Yanes A, Romero C, Brito de Azevedo E, Prada S, Martins A (2006) Los riesgos de las lluvias torrenciales en las islas de la Macaronesia (Azores, Madeira, Canarias y Cabo Verde). Clima Sociedad y Medio Ambiente 5:443–452

Masson DG (1996) Catastrophic collapse of the volcanic island of Hierro 15 ka ago and the history of landslides in the Canary Islands. Geology 24(3):231–234

Masson DG, Watts AB, Gee MJR, Urgeles R, Mitchell NC, Le Bas TP, Canals M (2002) Slope failures on the flanks of the western Canary Islands. Earth Sci Rev 57(1–2):1–35

Romero C (2016) Historias de volcanes. Isla de El Hierro. In: Arozena and Romero (eds) Temas y Lugares. Homenaje a Eduardo Martínez de Pisón. Serie Homenajes 7. Secretariado de Publicaciones. Universidad de La Laguna, pp 327–373

Romero C, Yanes A, Marzol V (2004) Caracterización y clasificación de las cuencas y redes hidrográficas en islas volcánicas atlánticas (Azores, Madeira, Canarias y Cabo Verde). IV Congrés Ibèric de Gestió y Planificació de LÁigua. Tortosa. Spain

Romero C, Yanes A, Marzol V (2006). Las áreas arreicas en la organización hídrica de las islas volcánicas atlánticas (Azores, Madeira, Canarias y Cabo Verde). In: Geomorfología y territorio: actas de la IX Reunión Nacional de Geomorfología, Santiago de Compostela, 13–15 de septiembre de 2006. Universidade de Santiago de Compostela, pp 697–710

Serrano E, Ruiz-Flaño P (2007) Geodiversity: a theoretical and applied concept. Geographica Helvetica 62(3):140–147

Strahler AN (1964) Quantitative geomorphology of drainage basins and channel networks. In: Chow BVT (ed) Handbook of applied hydrology. McGraw Hill Book Company, New York

Tibaldi A (1995) Morphology of pyroclastic cones and tectonics. J Geophys Res Solid Earth 100(B12):24521–24535

Geoheritage Inventory of the El Hierro UNESCO Global Geopark

Ramón Casillas Ruiz, Yurena Pérez Candelario, and Cristina Ferro Fernández

Abstract

In 2014 the Island of El Hierro (Canary Islands) was declared a Geopark of the Unesco network, thus becoming the first UNESCO geopark of the Canary Islands. The geological history of the Island of El Hierro can be understood through the visit of 61 geosites, which are representative of the growth and destruction of an oceanic Island in an intraplate environment. The geological heritage represented by these geosites has as foremost exponents those related to the formation of mega-landslides and the formation of extensive fields of pahoehoe lava-flows related to the historical or prehistoric fissure volcanism concerning the activity of its three rifts. This chapter describes the methodology used in establishing the geosite inventory carried out in 2019, as well as the description of the established geosites.

Keywords

Inventory • Geosite • Geopark • Geoheritage

1 Introduction

To proceed with the conservation and sustainable use of the geological heritage of any space, it is necessary to carry out the inventory of geosites (Carcavilla et al. 2009; García-Cortés et al. 2014; Brilha 2016). The Island Council of El Hierro presented the candidacy to be part of the "European Geoparks Network" (EGN) and "The Global Geoparks Network" (GGN) on October 1, 2012. In September 2014, the island of El Hierro it joins the EGN and GGN under the auspices of UNESCO, definitively declaring itself El Hierro Geopark. In November 2015, during the 38th General Conference of UNESCO, the approval of the "International Program of Earth Sciences and Geoparks (PIGG)" was ratified, thus entering El Hierro Geopark to form part of the World Geoparks Network of UNESCO, now renamed the UNESCO World Geopark.

The Geopark project Dossier contained a list and description of the Geosites and Geozones, which included a total of 7 Terrestrial Geozones, 3 Marine Geozones, 28 Terrestrial Geosites and 15 Marine Geosites.

Among the LIGs selected on that occasion was The Golfo Valley. The Golfo Valley is included in the Global Geosite VC010 "The El Golfo landslide (El Hierro)". It is a part of the Global Geosites inventory for Spain (Garcia-Cortés et al. 2008), a global inventory of the Earth's geological heritage (IUGS project had the support of ProGEO, IUCN and UNESCO). It is representative of the "Volcanic morphologies and edifices from the Canary Islands" nº 14 geological framework for Spain (Barrera 2009).

The review of the inventory of geosites of El Hierro Geopark, shown in this chapter, is a consequence of the need to face a precise identification, description and interpretation of the component elements of the island's geological heritage. These are understood as the whole of natural resources originated by geological processes and with scientific, cultural and/or educational value, such as the geological formations and structures, landforms, minerals, rocks, fossils, soils and other geological manifestations. They allow us to know, study and interpret the origin and evolution of the island of El Hierro, the processes that have shaped it, and the climates and landscapes of the past and present.

R. Casillas Ruiz (✉)
Geoparque de El Hierro-Departamento de Biología Animal, Edafología Y Geología de La Universidad de La Laguna-Cátedra Cultural "Telesforo Bravo" de La Universidad de La Laguna, San Cristóbal de La Laguna, Spain
e-mail: rcasilla@ull.edu.es

Y. Pérez Candelario
Reserva de La Biosfera Y Geoparque de El Hierro. Cabildo Insular de El Hierro, Santa Cruz de Tenerife, Spain
e-mail: yperez@elhierro.es

C. Ferro Fernández
Tourism and Innovation Freelance Consultant, Santa Cruz de Tenerife, Spain

J. Dóniz-Páez and N. M. Pérez (eds.), *El Hierro Island Global Geopark*, Geoheritage, Geoparks and Geotourism, https://doi.org/10.1007/978-3-031-07289-5_4

Thus, the present study starts from a double objective. The first one sicks to reflect the representativeness and the totality of the geodiversity of the island of El Hierro, which is barely formally characterized. The second is to identify the land and marine geosites of El Hierro Geopark.

To carry out this inventory of geosites of the El Hierro Geopark, there is a protocol already established worldwide that starts from the collection of information from the opinions of a panel of experts so that the points are selected according to their scientific values, educational, informative, tourist attraction, etc.

Therefore, this chapter will address the following aspects related to the recent Inventory of geosites of the El Hierro Geopark:

- The description of the methodology used in the establishment of geosites.
- The denomination of terrestrial land and marine geosites in El Hierro Geopark.
- The description of the proposal sheets for a geosite.
- The classification of the different geosites.

2 Methodology

To carry out the preliminary selection of geosites, the methodology proposed by the Spanish Geological Survey (IGME) (García-Cortés et al. 2000, 2014) has been followed.

2.1 Bibliographic and Documentary Compilation

The first task faced by the work team, made up of professors from the Departmental Unit of Geology of the Department of Animal Biology, Edaphology and Geology of La Laguna University, was the bibliographic and documentary compilation on the Geology of El Hierro. The information to be collected focused on four fundamental themes:

- Information available on the geological characteristics (with a multi-disciplinary nature) of the Island of El Hierro, and its geodynamic evolution. This information has included the MAGNA geological cartography and has served to become aware of the geosites that should be represented in the inventory and select the team of scientific collaborators who have been invited to participate in the selection of these geosites.
- Information on protected natural spaces and other elements of interest, both natural and historical and/or cultural heritage, as well as the legal regulations relating to of them. Its interest lies in knowing, on the one hand, what the level of protection of the elements to be inventoried can be and, on the other, what non-geological values can reinforce or complement the interest of the inventoried elements.
- Possible pre-existing geosite inventory initiatives in El Hierro Geopark. The work carried out rigorously in this field has been taken advantage of (such as the previous geosite inventory carried out in 2014, or the geosites collected in the reports of the geological sheets at 1:25,000 scale of La Restinga, Sabinosa, Valverde and Frontera).
- Guidebooks for scientific excursions and congresses carried out on the Island of El Hierro, such as the Geo-guides published by the Geological Society of Spain, as well as other guides on nature or protected natural areas that have sufficient scientific rigour.

2.2 Constitution of the Working Group and Election of Collaborating Experts

Given the complexity and variety of the geological record of the El Hierro UGG, both in time and space, it is easy to understand the need for expert collaborators in the different branches of Geology. Those support the inventory work team when selecting the most representative places for each of the themes involved in the geological diversity of the Island of El Hierro (Volcanology, Petrology-Geochemistry, Geomorphology, Sedimentology, Tectonics, Hydrogeology, Paleontology and Edaphology). Therefore, it is necessary to have experts who cover all these disciplines. The coordinating team selected these expert collaborators after analyzing the bibliography referring to the geology of El Hierro and invited them to participate in the inventory project.

2.3 Selection of the Geosites of the El Hierro UGG

To carry out a preliminary selection of all those places that, in the opinion of the work team and the expert collaborators, had the possibility of being incorporated into the inventory, we proceeded, in a similar way to that proposed by the Delphi methodology, described by García-Cortés et al. (2014), through several rounds of surveys carried out by all the experts.

Through these surveys, the experts were informed that they would carry out their geosites proposal taking into account the intrinsic values, those linked to their potential for use (scientific, educational or touristic) and those linked to their need for protection, following the proposal of Cendrero (1996), such as scientific knowledge, representativeness, rarity, type or locality of reference, state of conservation, protection status and legislation, conditions for the observation, geological diversity, scenery, scientific-didactic-touristic content and use, and presence of other natural or cultural assets.

3 Inventory of Geosites in the El Hierro UGG

Table 1 shows the geosites proposed for the El Hierro UGG in this report, classified according to the geological contexts defined on the Island of El Hierro by the coordinating team (shield vulcanism, rift vulcanism, central volcanic complexes; prehistoric and historical vulcanism; alluvial and fluvial-torrential processes and deposits; gravitational processes and deposits; coastal processes and deposits; geological elements submerged below sea level; volcanic or sedimentary aquifers; paleontological sites; tectonic structures; soils). This table also shows the main geological interests for each geosite. The location of these geosites appears in Fig. 1.

The table shows the geosites grouped into:

(A) The 18 geosites representing the 33% with the highest score in the opinion of the experts consulted (shown in blue in Fig. 1).
(B) The 29 geosites that obtained a score higher than 5, in the opinion of the experts consulted (shown in yellow in Fig. 1).
(C) The two remaining geosites are those that the coordinating team, in light of the regional knowledge of the geology of the Island of El Hierro, included in the best-valued 33%. However, they would not have deserved the recognition of the experts consulted (shown in black in Fig. 1).

Nevertheless, althogh the experts consulted only proposed a submarine geosite (EH-012. Tagoro Submarine Volcano), the coordinating team considered it necessary to include the submarine geosites that were already catalogued as geosites in the El Hierro Geopark in 2014 (shown in blue in Fig. 1): EH-048. El Salto; EH-049. El Diablo Cave; EH-050. El

Table 1 Classification of the main geological frameworks identified in El Hierro UGG

Geological framework of El Hierro UGG	Main geological interest representative of each geological framework	Code	Geosite denomination
(1) Shield volcanism	Vulcanological (Vul)	EH-001	El Golfo Valley. (6). (Geo)
		EH-003	Las Playas Valley. (6). (Geo)
		EH-009	Remains of the Hoya del Verodal tuff ring
		EH-021	The pyroclastic cones dissected from the cliffs (La Punta de los Reyes)
		EH-022	El Julan. (6). (Geo)
		EH-024	Ventejís Volcano. (Geo)
		EH-036	Lava-flows loaded with xenoliths from La Caleta. (Pe-Ge)
		EH-037	The trachyte lava-flow of the El Golfo Volcanic Edifice. (Pe-Ge)
	Geomorphological (Geo)	EH-034	The dyke of Jinama Landview
(2) Rift volcanism	Vulcanological (Vul)	EH-011	La Hoya de Fireba. (Geo)
		EH-014	The Montaña del Tesoro volcano, its lava-flows and the Tamaduste lava platform. (Geo)
		EH-015	El Pozo de la Calcosas. (7). (Geo)
		EH-025	La Caldereta. (Geo)
		EH-026	El Juaclo de las Moleras. (10). (Geo-Pal)
		EH-028	The field of volcanoes of the Suthern Ridge. (4). (Geo)
		EH-031	The Cala de Tacorón. (Geo)
		EH-041	The montaña de Puerto Naos. (Geo)
		EH-043	The Pico de la Mata cave. (10). (Geo-Pal)
		EH-044	The Curascán cave. (10). (Geo-Pal)

(continued)

Table 1 (continued)

Geological framework of El Hierro UGG	Main geological interest representative of each geological framework	Code	Geosite denomination
(3) Central volcanic complexes	Vulcanological (Vul)	EH-007	The Tanganasoga volcano. (Geo, Pe-Ge)
		EH-019	The Malpaso salic deposits. (Pe-Ge)
(4) Prehistoric and historical vulcanism	Vulcanological (Vul)	EH-002	El Lajial. (2). (Pe-Ge)
		EH-004	The Don Justo cave. (2). (Geo)
		EH-005	Orchilla volcanic group-eruptive fissures. Cliffs. (2). (Geo)
		EH-012	The Tagoro Submarine volcano. (2, 8)
		EH-023	The Lomo Negro volcano. (2)
(5) Alluvial and fluvial-torrential processes and deposits	Sedimentological (Se)	EH-013	The Fuga de Gorreta. (Geo)
		EH-045	The colluviums of the Barranco de las Arenas. (Pal)
	Geomorphological (Geo)	EH-046	Los Jables
(6) Gravitational processes and deposits	Sedimentological (Se)	EH-017	The debris-avalanche deposits from the 2nd gravitational slide responsible for the El Golfo valley formation
(7) Littoral processes and deposits	Vulcanological (Vul)	EH-015	The Pozo de la Calcosas. (2). (Geo)
		EH-029	Pillow-lava and hyaloclastic rocks at the base of the Tiñor Edifice in Timijirate. (1, 10). (Se, Pal)
	Geomorphological (Geo)	EH-020	The Roques de Salmor. (1). (Vol)
		EH-030	La Maceta. (4)
		EH-033	The Roque de la Bonanza. (1)
		EH-038	Coastal stone arches (Puntas de Gutiérrez). (4)
		EH-040	Coastal columnar joints in the Cachopo area. (4)
	Sedimentological (Se)	EH-027	The Arenas Blancas paleobeach. (4). (Vol, Pal)
		EH-032	The Verodal beach. (4). (Vol)
		EH-039	La Caleta paleobeach. (4). (Vol, Pal)
(8) Geological elements submerged below sea level	Geomorphological (Geo)	EH-048	El Salto. (2, 7). (Vol)
		EH-049	El Diablo Cave. (2, 7). (Vol)
		EH-050	El Bajón. (2, 7). (Vol)
		EH-051	Baja Bocarones. (1, 7). (Vol)
		EH-052	El Arco. (2, 7). (Vol)
		EH-053	La Hoya. (2, 7). (Vol)
		EH-054	Baja de la Palometa. (2, 7). (Vol)
		EH-055	El Charco Manso. (Vol). (2, 7). (Vol)
		EH-056	La Caleta. (2, 7). (Vol)
		EH-057	El Bajón del Puerto. (1, 7). (Vol)
		EH-058	El Roque de la Bonanza. (1, 7). (Vol)
		EH-059	La Baja de Anacón. (2, 7). (Vol)
		EH-060	Los Negros. (2, 7). (Vol)
		EH-061	El Barbudo. (2, 7). (Vol)
(9) Volcanic or sedimentary aquifers	Hydrogeological (Hy)	EH-042	The Garoé. (1). (Vol)
(10) Paleontological sites	Paleontological (Pa)	EH-008	The log molds of the Montaña Chamuscada lava flows. (4). (Vol)

(continued)

Table 1 (continued)

Geological framework of El Hierro UGG	Main geological interest representative of each geological framework	Code	Geosite denomination
(11) Tectonic	Tectonical (Tec)	EH-006	The San Andrés fault (in the Barranco de Tiñor). (6). (Geo)
		EH-016	The San Andrés fault (in the road to Puerto de la Estaca, in Morro del Jayo). (6). (Geo)
		EH-035	Antithetical faults of the Barranco de las Playecitas. (6). (Geo)
		EH-047	The graben between San Andrés-Tiñor road and La Cumbrecita. (6). (Geo)
(12) Edaphic	Edafological (Eda)	EH-010	El Jorado
		EH-18	Jondana

It includes all geosites of all geological frameworks. Secondaries geological frameworks and secondary main geological interests are also indicated in each geosite, in parentheses

Fig. 1 Location and distribution of the selected Geosites in the El Hierro UGG (the points and areas with a red transparent grid)

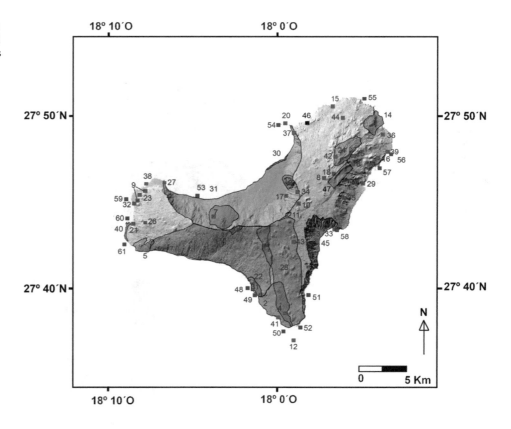

Bajón; EH-051. Baja Bocarones; EH-052. El Arco; EH-053. La Hoya; EH-054. Baja de la Palometa; EH-055. Charco Manso; EH-056. La Caleta; EH-057. Bajón del Puerto; EH-058. Roque de la Bonanza; EH-059. Baja de Anacón; EH-060. Los Negros and EH-061. El Barbudo.

After establishing the geosites of the Geopark and following the Methodology proposed by the IGME (García-Cortés et al. 2000, 2014), the team coordinating this review prepared the files for the Inventory of Geosites of El

Hierro UGG. A descriptive sheet has been made for each geosite (Table 2). This descriptive sheet includes the following aspects.

(A) The denomination of the geosite. In this case, two letters and three figures are used for the code (EH, El Hierro; 01, 2-digit code) and a name that describes the geological element and its geographical location (example: San Andrés fault in Barranco de Tiñor).

Table 2 Example of technical sheet of the inventory of geosites of El Hierro UGG

Proposal form for a geosite[a]				
Name of the geosite	EH-015. Pozo de la Calcosas			
Short description	Waterfall and lava delta formed by the arrival of lava-flows from the eruptive center of Montaña Aguarijo			
Justification of Interest	The Pozo de Las Calcosas is an excellent example of how a ravine can channel and advance the lava flows that jump a previous cliff, which fossilizes occasionally, and gain ground from the sea. It is a coastal area on cordate pahoehoe lavas that comes from two eruptive centers (Montaña Aguarijo and La Atalaya) and is located at a higher altitude. Currently, the area is affected by torrential processes, the dynamics of the slope with risks of landslides from the rocky edges of the cliff and by the action of the sea that generates interesting examples of abrasion platforms and "roques". Geomorphological and petrological interest			

Parameters justifying the choice of the geosite (mark with a cross those that you have considered)

☒ Representativeness	☒ Scenery
☒ Character of type or reference locality	☒ Informative content/informative use
☒ Scientific knowledge	☒ Didactic content/didactic use
☒ Conservation status	☐ Potential for recreational and outdoor activities
☒ Viewing conditions	☒ Links with other natural or cultural assets
☒ Rarity	☒ Geological diversity

Location	Province: Santa Cruz de Tenerife		Municipality (s): Valverde	
	Spot(s): (s) Pozo de Las Calcosas			
	Coordinates UTM[b]	**X: 20,975,690 E**	**Y: 308,302,450 N**	**Spindle: 28**
				Datum: REGCAN95
	In the event of being advisable to maintain the **confidentiality** of the site, by concealing its coordinates, please mark it with a cross (x)			☐
Access itinerary description	From Valverde, take the HI-5 towards Frontera, then take the HI-100 until the Pozo de Las Calcosas			
Situation diagram with delimitation proposal[c] (insert or attach fragment of map or SIGPAC orthophoto in a separate file)				

(continued)

Table 2 (continued)

Proposal form for a geosite[a]

Photograph (s) of the place (can be attached in separate files)	 Photo 1. Lava delta of the Pozo de Las Calcosas.  Photo 2. Detail of Photograph 1. Observe the arcs produced in the surface of the lava flow as it progresses.
Bibliographic references	• Becerril, L. (2014). Volcano-structural study and long-term volcanic hazard assessment on El Hierro Island (Canary Islands) (Ph.D. thesis document). University of Zaragoza, Spain. ISBN: 978-84-617-3444-3 • Carracedo, J. (2008). Los volcanes de las Islas Canarias (IV. La Palma, La Gomera, El Hierro). Ed. Rueda, Madrid. 213 pp • Carracedo, J. C. (2011). Geología de Canarias I (Origen, evolución, edad y volcanismo). Editorial Rueda S. L. • Carracedo, J. C.; Badiola, E. R.; Guillou, H.; De La Nuez, J. y Pérez Torrado, F. J. (2001). Geology and volcanology of La Palma and El Hierro, Western Canaries. Estudios Geológicos 57: 175–273 • González, E.; Dóniz-Páez, J.; Becerra-Ramírez, R.; Escobar, E.; Gosálvez, R. y Becerra-Ramírez, M. C. (2015). Itinerarios didácticos y geopatrimoniales por la isla de El Hierro. Ed. GEOVOL-UCLM, Ciudad Real, e-book, 272 p • IGME. (2010). Mapa y Memoria explicativa de la Hoja de Valverde (1105-II) del Mapa Geológico Nacional a escala 1:25.000 • Pellicer, M. J. (1975). Estudio vulcanológico, petrológico y geoquímico de la isla de El Hierro (Archipiélago Canario). Tesis Doctoral, Facultad de Ciencias Geológicas, Universidad Complutense de Madrid: 179 pp • Pellicer, M. J. (1977). Estudio volcanológico de la Isla de El Hierro, Islas Canarias. Estud Geol 33:181–197
Author of the proposal	Julio de la Nuez Pestana, Francisco Javier Dóniz Páez, José Luis Fernández Turiel, Laura Becerril Carretero, Francisco Javier Pérez Torrado, Alejandro Rodríguez González, y Ramón Casillas Ruiz

[a] The data provided will be treated as proposals that may be modified in later phases of the inventory
[b] From the geometric center of the place of geological interest
[c] Optional delimitation

Fig. 2 Examples of some geosites from the El Hierro UGG inventory: pahoehoe lavas from the EH-002 El Lajial geosite (**a**); San Andrés fault slickenside from the EH-006 The San Andrés fault (in the Barranco de Tiñor) geosite (**b**); the pyroclastic cones dissected by the cliff from the EH-021 The pyroclastic cones dissected from the cliffs (La Punta de los Reyes) geosite (**c**); and spectacular coastal caves with columnar joints from EH-040 Coastal columnar joints in the Cachopo área (**d**)

(B) Short description. The description of the rocky outcrop of interest is introduced in this section, providing the essential geosite data (lithology, structure, age, etc.).

(C) Justification of interest. In this section, it is necessary to indicate the interest rate of its content from the geologycal point of view: volcanological, petrological, tectonic, etc. In addition, evaluative comments are also introduced about the importance of the geosite concerning the interpretation of the geological history of the El Hierro, the exclusivity of the geosite and its relationship with other aspects of the heritage (historical, archaeological, ethnographic, etc.), as well as its valuation from the informative, didactic or recreational point of view.

(D) Justifying parameters of the choice of the place. This section refers to the evaluation parameters that we indicate in Sect. 2.3: the intrinsic values, the values linked to their potential for use (scientific, educational or touristic) and the values linked to their need for protection. Each geosite was assigned specific valuation parameters.

(E) Location. The province, the municipality and the toponymic name of the place where the geosite is located are indicated. The UTM coordinates of the center of the area encompassed by the geosite are also determined.

(F) Description of the access route. This section describes the path or route to be followed to visit the geosite. The names of the arrival roads or the approaching paths are identified, etc.

(G) Situation diagram with delimitation proposal. A map or an aerial or satellite photo with the delimitation of the geosite is also added for a better location.

(H) Photograph (s) of the place. In this section, the photographs were deemed appropriate for better identification and description of the geosite.

(I) Bibliographic references. This section lists the books, geological maps, scientific articles, etc., which describe the characteristics of the geosite.

(J) Author of the proposal. This last chapter indicates the members of the panel of experts who have proposed this outcrop as a geosite in the survey launch phase, described in Sect. 2.3 of this work (Table 2).

Acknowledgements This study has been carried out within the framework of a Specific Agreement between Cabildo de El Hierro and La Laguna University. We thank Luisa Maria Anceaume Chinea for the steps taken to formalize this Specific Agreement. We also thank the members of the expert panel for their invaluable collaboration: J. A. Gómez Sainz de Aja, Pedro Agustín Padrón Padrón, Constantino Criado Hernández, Javier Dóniz Páez, Alejandro Rodríguez González, Carolina Castillo Ruiz, María Esther Martín González, Laura Becerril Carretero, Jorge Yepes Temiño, Francisco J. Pérez Torrado, Julio de la Nuez Pestana, José Luis Fernández Turiel and José María Morales de Francisco. We also thank the VOLTURMAC (MAC2/4.6c/298) Project for their support in preparing this chapter.

References

Barrera JL (2009) Volcanic edificies and morphologies of the Canary Islands. In: García-Cortés A, Águeda Villar J, Palacio Suárez-Valgrande J, Salvador González CI (eds) Spanish geological frameworks and geosites. An approach to Spanish geological heritage of international relevance. Publicaciones del Instituto Geológico y Minero de España (IGME), Madrid, pp 146–156

Brilha JB (2016) Inventory and quantitative assessment of geosites and geodiversity sites: a review. Geoheritage 8:119–134

Carcavilla L, Durán JJ, García-Cortés A, López-Martínez J (2009) Geological heritage and geoconservation in Spain: past, present, and future. Geoheritage 1:75–91

Cendrero A (1996) El patrimonio geológico. Ideas para su protección, conservación y utilización. MOPTMA. In: El Patrimonio Geológico. Bases para su valoración, protección, conservación y utilización. Ministerio de Obras Públicas, Transportes y Medio Ambiente, Madrid, pp 17–38

García-Cortés A, Rábano I, Locutura J, Bellido F, Fernández-Gianotti J, Martín-Serrano A, Quesada C, Barnolas A, Durán JJ (2000) Contextos Geológicos españoles de relevancia internacional: establecimiento, descripción y justificación según la metodología del proyecto Global Geosites de la IUGS. Bol Geol Min 111(6): 5–38

García-Cortés A, Carcavilla L, Díaz-Martínez E, Vegas J (2014) Documento metodológico para la elaboración del Inventario Español de Lugares de Interés Geológico (IELIG). Propuesta para la actualización metodológica. Versión 5/12/2014. Instituto Geológico y Minero de España, pp 1–64

The Vegetation Landscapes of a Oceanic Recent Volcanic Island

Esther Beltrán-Yanes and Isabel Esquivel-Sigut

The islands that underwater fire has raised above the waves, gradually become overgrown with vegetation, but oftenthese newly formed lands are torn apart by the action of thesame forces that made them emerge from the bottom of the oceans. Perhaps certain islets that today are no more than heaps of slagand volcanic ash were once as fertile as the hills of Tacoronte and El Sauzal.

Alexander von Humboldt, Voyage aux régions équinoxiales du Noveau Continent fait en 1799, 1800, 1801, 1802, 1803 et 1804. I. Paris, 1816, p.113

Abstract

The aim of this chapter is to characterise the vegetation landscapes of El Hierro's Geopark, highlighting the important role played by the island's volcanic morphology in the richness and diversity of its landscapes. To this end, some of its most representative vegetation landscapes have been selected at various spatial scales, recognising their main discontinuities and internal organisation, and identifying the integrated combinations of the geographical factors that determine them have been identified, with special interest in the volcanic morphostructural conditioning factors. This work has required photointerpretation of aerial images and consultation of the WMS (*Web Map Service*) of IdeCanarias, as well as field work for the preparation of vegetation profiles and floristic-physiognomic inventories. Active volcanic areas are distinguished by being some of the most dynamic types of landscape on the planet. In this sense, the study of the vegetation landscapes of the small island of El Hierro allows us to discover how volcanic morphogenesis can extraordinarily diversify island landscapes.

Keywords

Vegetation landscapes • Volcanic rift • Geography of the vegetation • Post-eruptive plant colonisation • El Hierro's Geopark

1 Introduction

From a geographical point of view, the study of landscapes focuses on the analysis of the physiognomy of a territory and on the explanation of the interrelated forms and elements that make it up. From this point of view, the term landscape is inherent to any territory, though depending on its main element—natural, agricultural, urban, etc.—the techniques of analysis and information sources consulted will vary, the results of which must always be interpreted from a common approach inspired by the geographical foundation. Moreover, the territorial organisation that characterises a landscape is distinguished by a hierarchical structure of interdependent spatial units organised according to the spatial scale of study. It is important, therefore, to emphasise that each territorial organisation of a landscape is unique. Indeed, no two landscapes are exactly alike because their spatial structures are always different (Arozena Concepción and Beltrán-Yanes 2001; Bertrand and Bertrand 2006; Martínez de Pisón 2009; de Bolós i Capdevila and Gómez Ortiz 2009).

If the term 'landscape' is added to the term 'vegetation', the relevance of vegetation in the territorial organisation of the physiognomy of a space is highlighted. Vegetation's role tends to be very important, except in very cold or very dry

E. Beltrán-Yanes (✉) · I. Esquivel-Sigut
Area of Physical Geography, Department of Geography and History, University of La Laguna, San Cristóbal de La Laguna, Spain
e-mail: estyanes@ull.edu.es

I. Esquivel-Sigut
e-mail: iesquive@ull.edu.es

© The Author(s) 2023
J. Dóniz-Páez and N. M. Pérez (eds.), *El Hierro Island Global Geopark*,
Geoheritage, Geoparks and Geotourism, https://doi.org/10.1007/978-3-031-07289-5_5

places, since, together with the relief, vegetation is the main element in landscape characterisation, both because it contributes to its formal configuration, and because it is the component that best synthesises the interactions between the inert and the living. From this perspective, the plant component is fundamental for the identification of predominantly natural landscapes. Furthermore, the study of landscapes' territorial structure at different scales allows us to understand the spatial dimension of the interrelated factors that condition the geography that sustain landscape appearance. In short, in Biogeography from the perspective of Geography, the interest in knowing the living beings is inseparable from the territories of which they form part, because from this approach their knowledge allows the characterisation of the singularity of the territories (Arozena Concepción 1992).

Based on this geographic speciality, the aim of this chapter is to characterise the vegetation landscapes of El Hierro's Geopark, highlighting the important role played by the island's volcanic morphology in the richness and diversity of its landscapes. This study, therefore, focuses on how the relief created by continuous volcanic activity has conditioned the vegetation landscapes that distinguish this Geopark. To this end, some of its most representative vegetation landscapes have been selected at various spatial scales, recognising their main discontinuities and internal organisation, and identifying the integrated combinations of the geographical factors that determine them have been identified, with special interest in the volcanic morphostructural conditioning factors. This work has required photointerpretation of aerial images and consultation of the WMS (*Web Map Service*) of IdeCanarias (OrtoExpress and hillshade, vegetation and geology maps) (https://www.idecanarias.es/), as well as field work for the preparation of vegetation profiles and floristic-physiognomic inventories. Vegetation mapping was also carried out using GIS in some of the selected volcanic areas.

2　Volcanic Relief as a Diversifier of Vegetation Landscapes in the Canary Islands

Traditionally, when we focus on the study of vegetation in the Canary Islands, the first thing that strikes us is that the islands stand out worldwide for their biodiversity are parte of one the most important biodiversity hotspots on the planet, the Mediterranean Basin, (Médail and Quézel 1997) , favoured by their location at a subtropical latitude (27° 37′ and 29° 25′ north latitude and 13° 20′ and 18° 10′ west longitude). The position of these oceanic islands between Mediterranean and Tropical worlds allows them to display a significant range of flora and vegetation types, from forests adapted to thermophilic and dry environments, such as the

juniper forests, to humid environments, such as the original laurisilva or monteverde, or at higher altitudes, with the extensive Canary Island pine forests. Alongside these forest communities that occupy the midlands of the higher islands, xerophytic scrubland also grows on the coast, such as the cardonales-tabaibales, in which the floral and physiognomic affinities with the vegetation of the nearby African continent are noteworthy. In addition, on the highest peaks of the islands of La Palma (2426 m above sea level) and Tenerife (3717 m above sea level), the summit scrub has dominant connections not only to Mediterranean mountains but also has certain physiognomics links to tropical mountains.

As already mentioned, the location of the Canary Islands contributes to this range of forest and shrub plant communities. However, there is another geographical conditioning factor that determines this variety of vegetation types, which is the mountainous nature of these volcanic islands. The presence of mountainous areas that reach or exceed 1500 m above sea level on most of the islands introduces a geographical factor that causes striking variations in bioclimatic conditions. Among the main climatic consequences are the circulation of the trade winds at these latitudes in the eastern Atlantic. This means that the slopes of the highest islands, exposed to these winds, between 600 and 1500 m above sea level, receive the highest rainfall. In this altitudinal range, the "sea of clouds" also occurs, which provides notable environmental humidity due to its frequency and a considerable water volume that contributes to the survival of exuberant vegetation in these dry subtropical latitudes. For this reason, the altitude and orientation of the islands-mountains give rise to multiple environmental contrasts ranging from warm and dry local climates on the coast, to temperate and very humid on the windward slopes, cool with little rainfall on the leeward peaks and cold with irregular snowfalls on the highest peak of Tenerife (Marzol 2000, 2001).

Therefore, if we focus on the study of vegetation landscapes, the diversity of flora and vegetation is accompanied by an even more surprising wealth of plant communities. This is the result of novel environmental conditions caused by the complex volcanic orography of the islands, and which, on some of them are notably amplified by the constant rejuvenation of the relief due to volcanic activity.

In this chapter, we will focus on one of these island-mountains, built by recent volcanic activity: the island of El Hierro. The island is the smallest and most oceanic of the Canary Islands, with its highest point at the Pico de Malpaso (1501 m above sea level). El Hierro is distinguished by having the most diverse vegetation landscapes on recent volcanic morphologies at different spatial scales in the archipelago. The variety of Canary Island volcanism in terms of forms, processes, eruptive materials and chronology, interrelated with the subtropical climatic conditions of the islands, gives rise to multiple changes in vegetation landscapes over a very small and irregular island surface area. In addition, over the last few centuries, human action has also altered the natural vegetation landscapes.

3 The Vegetation Landscapes of El Hierro on an Island Scale

As already mentioned, El Hierro is one of the most recent islands of the archipelago and a large part of its territory is covered by volcanic cones and lava flows. It has traditionally been considered a dry island due to its geological youth, however, this is due more to water availability than to climatic drought. The recent nature of the volcanoes and the high porosity of the eruptive materials have not contributed to significant underground water storage, which, together with the absence of impermeable soils, has not facilitated the emergence of natural springs. However, its orography and more westerly position provide it with a generally humid environment that links El Hierro to the climatic regime of the other higher western islands.

In order to characterise the vegetation landscapes of El Hierro, it is necessary to initially distinguish the basic elements that make up its "territorial vegetation mosaics" and which correspond to the most representative plant communities of the main natural ecosystems of this island. However, these vegetation types may present internal nuances in their floristic composition and structure depending on geographical factors on a more local scale. Their unequal location, extension, distribution and continuity in the territory, as well as the spatial relationships established between them, constitute an original geography of the vegetation that distinguishes El Hierro's Geopark.

3.1 El Cardonal-Tabaibal

The cardonal-tabaibal grows between sea level and approximately 450 m above sea level (Aguilera et al. 1994) and is associated, over the first metres, with coastal halophilous scrubland in predominantly saline environments. On the island's southern slope, this vegetation unit reaches higher altitudinal levels until it meets the juniper forest (Aguilera Klink et al. 1994), and on the northern slope it extends up to 300 m above sea level. This scrub has physiognomic adaptations to a warm local climate, with average temperatures that can exceed 18 °C, and receives annual rainfall of between 150 and 250 mm (Marzol 1988; Aguilera et al. 1994). Adaptations to this climate include the development of succulent tissues, reduction of leaf area and spinescence to counteract evaportranspiration (Pérez de Paz et al. 1981; Sánchez Pinto 2005) (Fig. 1a).

The cardón (*Euphorbia canariensis*) and the sweet tabaiba *(Euphorbia baslsamifera)* are the most representative species of this plant community. Associated with this scrubland is the bitter tabaibal, which forms ecotonic zones with the juniper groves and whose most distinctive species is the bitter tabaiba (*Euphorbia lamarckii*). This floristic individual has a short, thick main trunk that branches out abundantly, giving rise to globular crowns. Accompanying species include *Kleinia neriifolia* (verode), *Periploca laevigata* (cornical), *Rumex lunaria* (vinagrera) and *Schizogyne sericea* (salado), among others. The richness and cover of this plant community can vary according to humidity conditions and the effects of human activity.

3.2 The Juniper Grove

One of the best representations of the thermophilic forest of the Canary Islands are the juniper forests of El Hierro. This open sclerophyllous forest of Mediterranean affinity receives annual rainfall totals between 200 and 500 mm and thrives at average temperatures between 17 and 19 °C (Aguilera Klink et al. 1994; Fernández-Palacios et al. 2008). In El Golfo, in the north of the island, there is a humid juniper forest located between 400 and 600 m asl, below regular contact with the sea of clouds.

To the south, El Hierro's junipers have a more open distribution and a characteristic creeping habit (Fig. 1b), arising from the adaptation of *Juniperus turbinata* subsp. *canariensis* to the frequent action of the prevailing winds at the summit. In addition to juniper, the most common species are bitter tabaiba (*Euphorbia lamarckii*), verode (*Kleinia neriifolia*), tasaigo (*Rubia fruticosa*) and salado (*Schizogyne sericea)* (del Arco et al. 2006).

3.3 The Monteverde or Laurel Forest

Above the juniper forests and on the windward slopes, under the regular effect of trade wind clouds, the monteverde forest has developed thanks to the mild temperatures between 13 and 17 °C and rainfall of around 1000 mm (Fernández-Palacios 2009). However, these laurel forests do not show mature expressions (Santos Guerra 2000), and some of the most characteristic species of this forest community are missing. In El Hierro's monteverde, mostly thermophilic species dominate, such as *Apollonias barbujana* subsp. *barbujana* (barbuzano), *Arbutus canariensis* (madroño), *Erica canariensis* (heather), *Morella faya* (faya), *Picconia excelsa* (paloblanco) and *Viburnum rugosum* (follao). These species form dense, evergreen forests (Fig. 1c) (Pérez de Paz 1990; Guzmán Ojeda et al. 2007), although when faya and heather predominate in some areas, they are mainly replacement

Fig. 1 Examples of the most characteristic plant communities of the island and which form part of the vegetation landscapes of the Geopark: **a** Cardonal-tabaibal in La Galera; **b** Juniper forest (La Dehesa); **c** Fayal-brezal (Jinamar); **d** Pine forest (El Julan). *Source* The authors

forests resulting from anthropic action and degradation. An example of this latter expression of monteverde is the fayal de La LLanía, with a floristic composition and structure that is highly conditioned by traditional livestock exploitation (Arozena Concepción et al. 2017).

3.4 The Pine Forest

Finally, the third forest community in the Geopark is the Canary Island pine forest, which is made up of stands of natural and reforested pine forest. It is located on the summit and leeward slopes of the island and is located at between 800 and 1450 m above sea level. The pine forest grows in a drier and thermally varied environment with annual rainfall above 300 mm and average annual temperatures ranging from 11 to 19 °C (Arévalo and Fernández-Palacios 2009). It is a tall, monospecific forest composed of Canary Island pine (*Pinus canariensis*) that forms an open tree canopy, with a very poor understory floristically in which *Lotus campylocladus* ssp (corazoncillo), *Trifolium* ssp., *Micromeria* ssp., *Hyparrenia hirta* and, locally, *Echium aculeatum* (ajinajo) can be found (Fig. 1d).

As has already been pointed out, these plant communities make up the main units or 'tesserae' of the different geographies that spatially articulate El Hierro's landscapes. From this point of view, a first general observation of the island reveals an altitudinal organisation of the vegetation with internal changes produced by orientation to the frequent trade winds. In the coastal sectors, a unit of xerophytic

scrubland can be recognised, and as the altitude increases, there is a concentration of the forest mass up to the summit of the island, which is clearly crowned. In this forest unit, there are physiognomic and floristic differences between the thermophilus forest of the lower altitudinal areas, the monteverde forest of the higher midlands to the windward and on the summits, and the pine forests that extend along the culminating sectors and slopes on the leeward side.

This main territorial organisation of the vegetation landscape is due to the presence of a vigorous and abrupt volcanic relief with a maximum age of 1.2 Ma (Guillou et al. 1996), which at its highest point reaches 1501 m asl and is distinguished by a triangular layout, built from the triaxial system of high volcanic edifices or rifts that constitute its volcanic summits (Guillou et al. 1996; Carracedo et al. 2001). The volcanic rift is a longitudinal morphostructure characterised by volcanic activity, and whose construction is spatially organised around a main tectonic axis that accounts for the greatest number of eruptive phenomena of predominantly basaltic composition. A large volcanic construction is thus created by the superposition and juxtaposition of multiple monogenetic volcanoes (Romero 1986, 1991; Dóniz-Páez 2009). The spectacular topographic amphitheatre of El Golfo, located to the north of the island, and resulting from one of the most important gravity slides ever to have occurred on El Hierro (between 21,000 and 130,000 years ago; Carracedo 2008), gives it its characteristic crescent shape. The small surface area in relation to the maximum height of the island means that El Hierro has the steepest average slopes in the archipelago.

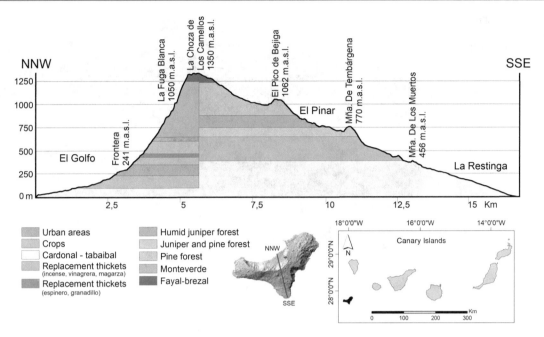

Fig. 2 NNW-SSE vegetation profile of the island of El Hierro. *Source* Idecanarias, Self-elaboration

Consequently, the altitude of El Hierro's volcanic relief explains the variety of forests it presents by hindering the circulation of oceanic winds, giving rise to environmental contrasts and the consequent discharge of abundant water resources on the windward slopes and on part of its summits. The vegetation profile that summarises this geography of the vegetation landscape runs from north to south through the central part of the Geopark (Fig. 2).

This profile begins on the coast of the Valle del Golfo, around Los Arenales, where the halophilic plant community of thyme (*Frankenia ericifolia*) and servilleta marina (*Astydamia latifolia*) grows and is in contact with the maresía, and the sweet tabaibales and cardonales typical of the coastal xerophytic environment. The extension of this part of El Golfo has facilitated greater settlement (Belguera, Tigaday, Los llanillos, etc.) and the expansion of agriculture, reinforced over the last few decades, which has profoundly altered the original geography and characteristics of the cardonales-tabaibales and the humid juniper groves. Indeed, this has left the latter plant community reduced to a small discontinuous unit at around 500 m above sea level. As we approach the escarpment of El Golfo and ascend, we can recognise different replacement thickets, such as bitter tabaibales (*Euphorbia lamarckii*), inciensales (*Artemisia thuscula*) and granadillares (*Hypericum canariense*) among the abandoned crops of the dominant detrital accumulations. However, it is from this last altitudinal level onwards, where the imposing escarpment of El Golfo, with slopes that reach a height of 1000 m in the easternmost sector, gives rise to a marked spatial change in the floristic composition and physiognomy of the vegetation.

The presence of this high wall that encloses this great depression to the south has further protected the forests from human activity and introduced a strict altitude control as an organiser of the forest landscape. The monteverde forest and fayal-brezal vegetation landscape units therefore appear territorially suspended vertically, as if they were literally ascending the slope until they reached its summit (Fig. 3).

This clear discontinuity in the landscape between scrubland and woodland in the north of the island is further reinforced by the concave and semi-circular topography of El Golfo to the NNW. This facilitates the concentration of water from the Atlantic winds, essential for the survival of the monteverde forest.

Once this great escarpment has been crossed, the vegetation landscape adapts to the presence of the N–S rift of El Pinar, which generates a volcanic alignment with a general convex topography that descends progressively to the southern tip of the island. The construction of this volcanic rift from parallel structural lines with a dominant north–south direction (Carracedo 2008) has given rise to a large longitudinal volcanic relief with a flat summit and a maximum height of approximately 1300 m above sea level.

The first thing that is striking about the vegetation on this southern slope is the spatial organisation of the forests. The flow of trade wind clouds over the summit allows the growth of fayal-brezal on the upper areas of this slope, with the pine forest (*Pinus canariensis*) below it. Only locally can this distribution be modified by the spatial discontinuities caused by the lapilli surfaces of the most recent eruptions or by the pine reforestation conducted on El Hierro during the last century (https://www.idecanarias.es/). Therefore, the pine

Fig. 3 Image of El Golfo with the dense forest cover of the monteverde on the escarpment. *Source* http://www. fotosaereasdecanarias.com

forest in its natural distribution is characterised by a forest floor on the leeward side of the slope, which currently covers large areas located between the escarpment of Las Playas and the slopes of El Julan.

As we go down this slope, the pine forest is replaced by juniper forest through an ecotonic floor. However, the flat summits of this volcanic morphostructure have been used for traditional livestock farming and agriculture, typical of the midlands of the south of the island with its water and soil resources, so that from around an altitude of 800 m, the forest is discontinuously located spatially. In this sector, areas of crops can be recognised, alternating with replacement scrubland in those plots where agriculture has ceased. Here, incense bushes (*Artemisia thuscula*) are frequent, together with other nitrophilous scrubland, which is also conditioned by grazing. Finally, from 400 to 350 m above sea level, there is a clear spatial discontinuity in the vegetation landscape, which is related to the presence of extensive, more recent volcanic surfaces corresponding to volcanic cones, badlands and coastal lajiales. This coastal sector has an open cover of dominant species, such as salado (*Schizogyne sericea*), also known on the island as irama, and tabaibas (*Euphorbia lamarckii*), in which the age of the volcanoes, the type of volcanic substrate and traditional farming uses are determining factors in interpreting the floristic changes and the structure of the vegetation.

4 The Vegetation Landscapes of the Volcanic Rifts of the Geopark

Differences in location, spatial arrangement and topography of El Hierro's rifts, originating from the recent volcanism, mean that the vegetation landscapes of these volcanic alignments present striking contrasts in their features and spatial organisation.

Thus, on the N-S southern volcanic rift, its location on the leeward side of the island gives the pine forest landscape a prominent role, whereas on the north-eastern volcanic rift, on the windward side, there is monteverde forest. This volcanic edifice, with an altitude of just over 1300 m above sea level and built from NE–SW oriented tectonic patterns (Carracedo 2008), has high humidity that promotes the development of monteverde forest at its summit, as can be seen today on some of its volcanic cones. However, much of the laurel (monteverde) forest of these mountains was ploughed up for agricultural use in the seventeenth century due to the quality of its volcanic soils (Hernández and Niebla 1985).

The exceptional conditions of humidity and the presence of this forest generated the best land on the island on a plateau, which has been given over to traditional livestock farming alternating with subsistence crops. Therefore, at present, on the north-eastern rift, an altitudinal organisation of the vegetation landscape can be identified in which there is a cardonal-tabaibal in different states of conservation on the coast, replaced from 250 to 300 m above sea level by shrub and nitrophilous formations in areas where traditional farming uses have ceased. The thickets of bitter tabaiba (*Euphorbia lamarckii*), vinagreras (*Rumex lunaria*) and incense (*Artemisia thuscula*) alternate in the landscape depending on altitude, orientation and the time when farming or agricultural activity ceased, until they meet at the summit with fayal-brezal of the protected area of Ventejís (Fig. 4). In this sector, the fayal-brezal is spatially combined with other vegetation units such as thistle grasslands (*Galactites tomentosus*) and tagasaste crops (*Chamaecytisus proliferus* subsp. *proliferus* var. *palmensis*), which highlight the use of this mountain for preferential livestock farming. The upper forest floor of this second profile corresponds to a reforested pine forest with the foreign species *Pinus radiata*, which replaces the monteverde characteristic of this volcanic rift.

Fig. 4 NE-SW vegetation profile (1) and NNW-SSE topographic profile of the culminate plain (2) on the northeast volcanic rift. *Source* Idecanarias, Self-elaboration

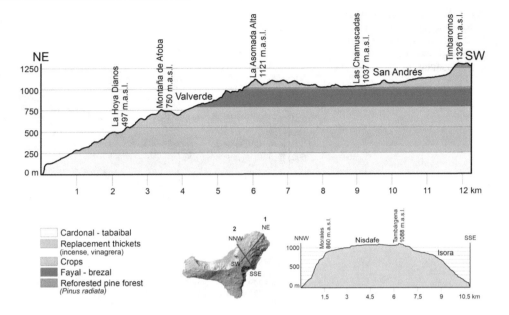

By contrast, the western orientation of the island's third volcanic rift with a WNW–ESE main orientation, together with its original local topography, causes striking differences in the Geopark's vegetational landscapes. This high alignment of recent volcanoes has its highest point on the island at Pico de Malpaso (1501 m above sea level). Additionally, it has the peculiarity that the large landslides of El Golfo to the north and Julan to the south have considerably narrowed its summit. On the other hand, the summit between approximately 400 and 1000 m above sea level has a flat and more extensive topography, which descends gently until it reaches a cliffy coastline.

The vegetation landscape of this volcanic rift is organised altitudinally with a xerophytic coastal scrubland dominated today by irama (*Schizogyne sericea*), which is replaced at around 350 m above sea level by a juniper forest in the La Dehesa area, which is one of the most representative forest landscapes on the island (Fig. 5). The characteristic aerodynamic shapes of this open forest are due to the persistent action of the NE winds (Fig. 1b) at the summit. The fact that this altitudinal section corresponds to an orographic sector of considerable extension emphasises the juniper forest in the vegetation landscape of this volcanic mountain, whose regeneration, biodiversity and structure are conditioned by past use of this forest (Salvá Catarineu et al. 2012). The vegetation profile of the summit of this volcanic rift is completed at altitude by the pine forests of *Pinus canariensis*, whose surface area has been enlarged by reforestation. Alongside these forests, replacement scrub and grasslands composed of bitter tabaiba (*Euphorbia lamarckii),* vinagreras (*Rumex lunaria*), incense (*Artemisia thuscula)* and thyme (*Micromeria hyssopifolia*) are organised spatially depending on altitude and orientation, and, once

again, they reflect common livestock use on the island. However, discontinuities in the vegetation landscape caused by recent volcanic events on this rift, such as those linked to the summit pyroclast fields of the Tanganasoga volcano, also locally organise the vegetation of this volcanic mountain alignment.

5 Vegetation Landscapes on a Local Scale: The Tesoro Volcano

The analysis of vegetation landscapes on a larger spatial scale allows us to identify new vegetation landscapes in these mountains, related to the most recent volcanism. This is a typical disturbance factor in the natural dynamics of vegetation landscapes in active volcanic territories. These landscapes respond to a process of primary plant succession in which primocolonising vegetation settles on a rocky surface devoid of soil, with very low fertility and lacking in organic matter (Smathers and Mueller-Dombois 1972; Hendrix 1981).

For the study of this type of vegetation landscape, the Tesoro volcano has been selected. It is located on the north-eastern volcanic rift, in Tamaduste, a coastal rift. This Holocene volcanic edifice of monogenic character and basaltic composition presents a volcanic cone with funnel craters and basal emission centres through which most lava of different typologies flowed (Dóniz-Paez et al. 2009). The overflow of the lava along the rift generated lava deltas from the superposition of lava flows, in which lava forms aa, lava balls and lava blocks can be identified. Likewise, in the interior of the volcano, some forms of modelling can also be recognised, such as detrital fans of torrential origin, which cover certain parts of the coastal lava platform.

Fig. 5 Vegetation profile (1) and topographic profile of the La Dehesa sector (2) in the western rift. *Source* Idecanarias, Self-elaboration

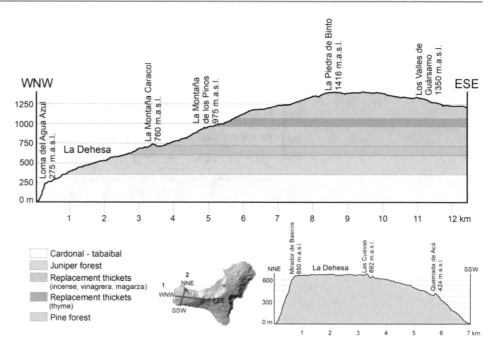

The climatic zone in which this volcano is located is semi-arid, characteristic of Canary Island coasts, although its exposure to the humid north-easterly winds means that it receives high levels of humidity most of the year. The vegetation is therefore defined by the presence of a xerophytic scrubland made up mainly of *Rumex lunaria, Kleinia neriifolia, Schizogyne sericea* and various species of the genus *Aeonium* sp. accompanied by a tapestry of thallophytes, which presents varied spatial units depending on the floristic composition, size and cover of the shrub formation.

When the vegetation landscapes of the most recent volcanoes are analysed in a semi-arid environment, which is the dominant one in the Canary Island archipelago, thus conserving the original volcanic forms hardly transformed by modelling and erosion, it is surprising to see the forceful control exerted by the morphology of the Tesoro volcano in the process of plant colonisation (Beltrán Yanes 2000; Beltrán Yanes and Dóniz Páez 2009). Topography, through local environmental changes and the regulating effects of morphoclimatic processes, as well as the shapes and surface textures of the volcanic substrates, constitute the fundamental geographical determinants of the vegetation landscape. In this way, different types of scrublands can be distinguished in El Tesoro that coincide territorially with its main morphological units and present the following characteristics (Fig. 6).

On the volcanic cone and lapilli fields, in addition to accumulations of pyroclasts and lava slopes (highly fragmented lava material) that cover the cliff, there are mainly calcareous plants (*Rumex lunaria*) that form an open scrubland, affected by the lack of stability of these eruptive materials. The topography of the volcanic mountain and the

cliff also accentuate the displacements of the substratum by gravity and torrentiality.

On the other hand, on the lava flows located on the cliff, and therefore in the same environmental sector as the previous unit, the more stable and continuous nature of the lava surfaces favours a significant presence of aerohygrophilous and heliophilous lichens, such as *Ramalina bourgaeana, Xanthoria resendei,* etc., which do not grow on the pyroclasts. Associated with the thallophytes, there is also open scrub, but more diverse, of *Kleinia neriifolia, Rumex lunaria, Aeonium* sp. and ferns, such as *Allosorus fragilis,* helped by these substrate characteristics.

On the coast, the lava flows show other local variations in vegetation, strongly influenced by the saline environment, but, above all, by the morphology of lava flows. In the thick blocky lava flows, there is a minimal presence of vegetation with isolated elements of *Kleinia neriifolia* with clear rupicolous and fissuriferous adaptations. These are very massive lavas with deep intercalated cracks that show very limited weathering processes. By contrast, on the aa lava flows located to the north of the lava delta, their more scoriaceous and vacuolar texture facilitates the disintegration of the lava substrate and, therefore, an increase in the presence of vegetation with a thicket of *Rumex lunaria, Schizogyne sericea and Kleinia neriifolia.*

However, one of the most striking vegetation units from the point of view of plant colonisation of this volcano are the dense thickets of *Schizogyne sericea* on the alluvial deposits on the lava platform. The development of detrital fans originating from the concentrated runoff through lava channels on the coastal slope has given rise to thickets with the highest cover rates on the volcano. This halophilic

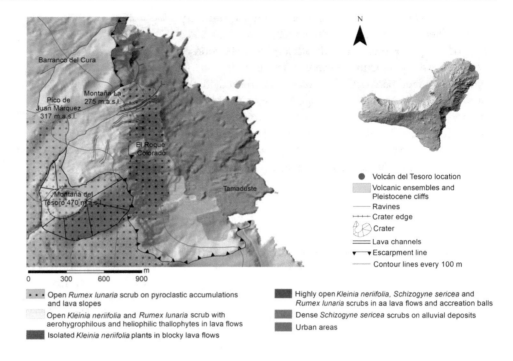

Fig. 6 Vegetation map of Volcán del Tesoro. *Source and cartographic base* Idecanarias. Self-elaboration

● Volcán del Tesoro location
▨ Volcanic ensembles and Pleistocene cliffs
── Ravines
┈┄ Crater edge
Crater
══ Lava channels
▼─▼ Escarpment line
── Contour lines every 100 m

• • • Open *Rumex lunaria* scrub on pyroclastic accumulations and lava slopes

Open *Kleinia neriifolia* and *Rumex lunaria* scrub with aerohygrophilous and heliophilic thallophytes in lava flows

Isolated *Kleinia neriifolia* plants in blocky lava flows

Highly open *Kleinia neriifolia, Schizogyne sericea* and *Rumex lunaria* scrubs in aa lava flows and accreation balls

Dense *Schizogyne sericea* scrubs on alluvial deposits

Urban areas

scrubland covers 60% of the surface with a maximum height of 1.50 m. The main floristic elements are salt cedar including *Schizogyne sericea , Rumex lunaria* and *Kleinia neriifolia.* The existence of this plant community constitutes a unique vegetation unit in the study of the plant colonisation of recent volcanic territories. It is associated with the allochthonous soils, which even contain their own biological capital (propagules and seeds), and are the result of a very rapid local transformation of the original morphology of the volcano (Beltrán Yanes 2000). From the perspective of the study of the influence of biotic and abiotic factors on the changes in time and space of post-eruptive colonising plant communities, there are interesting contributions made in other volcanic areas such as the Kula Volcano, Turkey (Öner and Oflas 1977), Paricutin (Mexico) (Velázquez et al. 2000) and the Tolbachinskii Dol Volcanic Plateau (Korablex and Neshataeva 2016) and Tolbachinsky Dol, in Kamchatka (Grishin 2010).

6 Conclusions

El Hierro is an excellent example of the close relationship between volcanic morphogenetic processes and the island's vegetation landscapes. The wealth of volcanic morphologies at various scales creates multiple, interdependent vegetation geographies that also contribute to defining the territorial physiognomic identity.

Thus, taking an overview, not only does the vigorous volcanic relief establish spatial contrasts in the vegetation between the windward and leeward sides of the island, but also the basic structure of its orographic framework, traced by the central crossroads of volcanic rifts with different morphology and spatial arrangement, establishes other variations in the bioclimatic conditions and vegetation landscapes. In this sense, the altitude and orientation of these mountains with respect to the trade winds determines the differentiated organisation of their plant mosaics. But also, from this perspective, it is striking how the different locations of the plateaus in these orographic elevations, sculpted by the gravitational landslides on their slopes, give rise to an unequal landscape relevance of the various forests in El Hierro. This novel volcanic relief has even conditioned the spatial organisation of the traditional uses and exploitation of the island's natural resources. Therefore, this factor is also essential for the geographical interpretation of the current state of the vegetation, reflected in the structure and floristic composition of the plant communities.

On a larger spatial scale, El Hierro's vegetation landscapes offer new characteristics and spatial structures derived from the most recent volcanic activity. This activity has introduced a factor of perturbation and renewal of the natural dynamics of the plant communities and landscapes. In these cases, the age of new structures and their morphological units are determining factors in the territorial change of the vegetation, although they are always dependent on the prevailing climatic conditions (Beltrán-Yanes 1992).

In short, active volcanic areas are distinguished by being some of the most dynamic types of landscape on the planet. In this sense, the study of the vegetation landscapes of the small island of El Hierro allows us to discover how volcanic morphogenesis can extraordinarily diversify island landscapes. There is no doubt that eruptions suddenly change, alter and destroy the vegetation, fauna and landscape of the places affected, with often catastrophic consequences for the population. However, this chapter aims to highlight another perspective of volcanism related to the renewal of landscapes and its important role in the construction of new and original territorial configurations, which, from the point of view of the study of the vegetation landscapes of oceanic volcanic islands, the Canary archipelago stand out worldwide for its diversity.

References

Aguilera Klink F, Brito Hernández A, Castilla Gutiérrez C, Díaz Hernández A, Fernández-Palacios JM, Rodríguez Rodríguez A, Sabaté Bel F, Sánchez García J (1994) Canarias. Economía, ecología y medio ambiente. Francisco Lemus, San Cristóbal de La Laguna

Arévalo JR, Fernández-Palacios JM (2009) 9550 Pinares endémicos canarios. In: VVAA, Bases ecológicas preliminares para la conservación de los tipos de hábitat de interés comunitario en España. Dirección General de Medio Natural y Política Forestal, Ministerio de Medio Ambiente, y Medio Rural y Marino, Madrid, 74 p

Arozena Concepción ME (1992) Consideraciones en torno al puesto de la Biogeografía en la Geografía. Alisios 2:22–34

Arozena Concepción ME, Beltrán Yanes E (2001) Los paisajes vegetales. In: Fernández-Palacios JM, Martín Esquivel JL (Dirs y coords) Naturaleza de las Islas Canarias. Ecología y Conservación. Turquesa Publicaciones, pp 95–102

Arozena Concepción ME, Panareda Clopés JM, Martín Febles VM (2017) Los paisajes de la laurisilva canaria. Editorial Kinnamon, Santa Cruz de Tenerife

Beltrán-Yanes E (1992) La vegetación como criterio para establecer la cronología de la actividad volcánica reciente en Tenerife (I. Canarias). Actas VI Coloquio Ibérico de Geografía, Porto, pp 795–799

Beltrán Yanes E (2000) El paisaje natural de los volcanes históricos de Tenerife. Fundación Canaria Mapfre-Guanarteme, Las Palmas de Gran Canaria

Beltrán Yanes E, Dóniz Páez J (2009) 8320 Campos de lava y excavaciones naturales. In: VVAA Bases ecológicas preliminares para la conservación de los tipos de hábitat de interés comunitario en España. Dirección General de Medio Natural y Medio Rural y Marino, Madrid, 124 p

Bertrand C, Betrand G (2006) Geografía del Medioambiente. Editorial Universidad de Granada, Granada

Carracedo JC, Rodríguez-Badiola E, Guillou H, Nuez Pestana JDL, Pérez Torrado FJ (2001) Geología y vulcanología de la Palma y el Hierro, oeste de Canarias. Estud Geol 57:175–273

Carracedo JC (2008) Los volcanes de las Islas Canarias IV. La Palma, La Gomera y El Hierro. Rueda, Madrid

del Arco M (ed) (2006) Memoria General. Mapa de Vegetación de Canarias. Litografía A. Romero, S.L. San Cristóbal de La Laguna

de Bolós i Capdevila M, Gómez Ortiz A (2009) La ciencia del paisaje. In: Busquets J, Cortina A (coords) Gestión del paisaje. Editorial Ariel, Barcelona, pp 165–180

Dóniz-Páez J (2009) Volcanes basálticos monogenéticos de Tenerife. Concejalía de Medioambiente del Excmo. Ayuntamiento de Los Realejos

Dóniz Páez J, Beltrán Yanes E, Romero Ruiz C (2009) Unidades geomorfológicas y de paisaje del litoral volcánico de El Tamaduste (El Hierro, Islas Canarias, España). XXI Congreso de Geógrafos Españoles, Ciudad Real

Fernández-Palacios JM, Otto R, Delgado JD, Arévalo J R, Naranjo A, Gónzalez Artiles F, Morici C, Barone R (2008) Los bosques termófilos de canarias. Proyecto LIFE04/NAT/ES/000064. Cabildo Insular de Tenerife, Santa Cruz de Tenerife

Fernández-Palacios JM (2009) 9360 Laurisilvas macaronésicas (Laurus, Ocotea)(*). In: VVAA, Bases ecológicas preliminares para la conservación de los tipos de hábitat de interés comunitario en España. Dirección General de Medio Natural y Política Forestal, Ministerio de Medio Ambiente, y Medio Rural y Marino, Madrid, 68 p

Grishin SY (2010) Vegetation changes under the impact of volcanic Ashfall (Tolbachinsky Dol, Kamchatka). Russ J Ecol 41(5):436–439

Guillou H, Carracedo JC, Pérez Torrado F, Rodríguez Badiola E (1996) K–Ar ages and magnetic stratigraphy of hotspot-induced, fast grown oceanic island: El Hierro, Canary Islands. J Volcanol Geoth Res 73(1–2):141–155

Gúzman Ojeda J, Cabrera Calixto F, Melián Quintana A (2007) Árboles de Canarias. Guía de campo. Gobierno de Canarias, Gran Canaria

Hernández Hernández J, Niebla Tomé E (1985) El Hierro. In: Afonso L (Dirs) Geografía de Canarias Tomo 4. Editorial Interinsular Canaria, Santa Cruz de Tenerife, pp 146–180

Hendrix LB (1981) Post-eruption succession on isla Fernandina. Galápagos. Madroño 28(4):242–254

Infraestructura de Datos Espaciales de Canarias (IDECanarias). https:// www.idecanarias.es/

Korablex AP, Neshataeva VY (2016) Primary plant successions of forest belt vegetation on the Tolbachinskii Dol Volcanic Plateau (Kamchatka). Biol Bull 43(4):307–317

Martínez de Pisón E (2009) Los paisajes de los geógrafos. Revista Geographicalia 55:5–25

Marzol Jaen V (1988) La lluvia, un recurso natural para Canarias. Servicio de Publicaciones de la Caja General de Ahorros de Canarias, Santa Cruz de Tenerife

Marzol Jaén MV (2000) El Clima. In: Morales G, Pérez R (Dirs y coords) Gran Atlas Temático de Canarias. Editorial Interinsular Canaria, Santa Cruz de Tenerife, pp 87–106

Marzol Jaén MV (2001) Los factores atmosféricos y geográficos que definen el clima del archipiélago canario. In: Raso Nadal JM (ed) Proyectos y métodos actuales en climatología (conferencias invitadas al I Congreso de la AEC). Asociación Española de Climatología, pp 151–176

Médail F, and Quézel P, (1997) Hot-spots analysis for conservation of plant biodiversity in the Mediterranean Basin. Ann Mo Bot Gard 84:112–127

Öner M, Oflas S (1977) Plant succession on the Kula Volcano in Turkey plant. Ecology 34(1):436–439

Pérez de Paz L, del Arco M, Wildpret W (1981) Contribución al conocimiento de la flora y vegetación de El Hierro (Islas Canarias). Lagascalia 1:25–57

Pérez de Paz P (1990) Parque Nacional de Garajonay. ICONA, Madrid

Romero C (1986) Aproximación a la sistemática de las estructuras volcánicas complejas de las Islas Canarias. Eria 11:211–223

Romero C (1991) Las manifestaciones volcánicas históricas del Archipiélago Canario. Consejería de Política Territorial. Gobierno de Canaria, 2 tomos

Salvá Catarineu M, Romo A, Salvador Franch F (2012) Estructura de edad y biodiversidad de los sabinares de Juniperus turbinata Guss. en El Hierro (Islas Canarias). VII Congreso Español de Biogeografía, Pirineo, Sant Pere de Ribes

Sánchez Pinto L (2005) Las euforbias de Canarias. Rincones Del Atlántico 2:60–65

Santos Guerra A (2000) La Vegetación. In: Morales Matos G, Pérez González R (eds) Gran Atlas Temático de Canarias. Editorial Interinsular Canarias, S.A. Santa Cruz de Tenerife, pp 121–145

Smathers GA, Mueller-Dombois D (1972) Invasion and recovery of vegetation after volcanic eruption in Hawaii. Honolulu. International biological program technical report, p 10

Velázquez A, Gimenez de Azcárate J, Gerardo B, Escamilla M (2000) Vegetation dynamics on Paricutín recent mexican volcano Landscapes. Acta Phytogeographica Suecica 85:71–78

Human Occupation of a Small Volcanic Island

Carlos S. Martín Fernández

Abstract

El Hierro has had an evolution in terms of its population in close relation to a series of historical, economic and social transformations that have left an important mark on the landscape. In general terms, we could affirm that the evolution and distribution of the population has been conditioned by economic cycles and modes of production, which have worked in accordance with the interest of local and regional elites.

Keywords

Population • Economy and society

1 Pre-European Population and Settlement

The first *Bimbache* or *Bimbape* (names given to indigenous people) settlement on the island of El Hierro is based on the speculative and hypothetical assumptions that characterize the scientific studies of the Canary Islands' prehistory. Thus, despite the advances in archaeological science, there are still more questions than answers regarding the islands' early pre-European settlers.

We can, however, affirm the North African character of El Hierro's first settlers, without specifying how and why they settled on the island. The first settlement was around 338 A.D. with a possible date range of between 212 A.D. and 489 A.D. (Jiménez Gómez 1993). This population survived by rearing small livestock, rudimentary cereal cultivation and other farming and fishing activities, all in an environmental context characterized by a small territory and limited water and soil resources. These natural circumstances together with the pirate raids in search of slaves in the years prior to their conquest were a serious limitation for El Hierro's population development.

Determining the exact population size for this period is a challenging archaeological task. While awaiting advances in the field, ethnohistorical sources can, however, provide us with some data, starting with the chronicles of the friars who accompanied the first expeditions to conquer the islands at the beginning of the fifteenth century led by Jean de Béthencourt and Gadifer de La Salle. This source, for a period immediately prior to the first European invasion, limits itself to saying that on El Hierro "few people remain" (Le Canarien 1959), in clear reference to the abundant slave raids to which the island had been subjected for a long time.

But how many people could we be talking about? In the absence of exact sources, estimates point to a potential *Bimbache* population prior to the conquest of between 500 and 1400 inhabitants (Macías Hernández 1992; Junyent 2013), who were organized in dispersed villages without constituting permanent settlements, taking advantage of the different and seasonal bioclimatic periods to obtain pastures and access to water. From very early on, a fundamental characteristic of these settlements was their periodic vertical and horizontal seasonal mobility.

2 The European Occupation

From 1404 to 1405, after the arrival of the troops of Béthencourt and La Salle, the first permanent settlement of Europeans on El Hierro occurred.[1] It was made up of 120

C. S. Martín Fernández (✉)
Department of Geography and History, University of La Laguna, San Cristóbal de La Laguna, Spain
e-mail: csmartin@ull.edu.es

---

[1] The conquest and colonization of the island of El Hierro was carried out in two phases: the first by Norman nobles and the second by Castilian nobles, both in the service of the Kings of Castile. This form of conquest and colonization is known as seigniorial and is characterized as a particular enterprise of the lords, who obtained feudal or feudal rights from the king over the conquered lands. El Hierro and La Gomera shared a lordship with a hereditary character until its disappearance in the nineteenth century.

© The Author(s) 2023
J. Dóniz-Páez and N. M. Pérez (eds.), *El Hierro Island Global Geopark*, Geoheritage, Geoparks and Geotourism, https://doi.org/10.1007/978-3-031-07289-5_6

inhabitants (Le Canarien 1959), French and Flemish, to whom López de Ulloa in 1646 also added Castilians and indigenous people from the neighbouring island of La Gomera (Morales 1978).

During this first stage, also called the Norman stage due to the origin of the conquerors, to avoid the departure of the small number of settlers, the Lord exempted them from paying rents and gave them land and caves to give them a fair chance of survival. In an extensive and imprecise area, this first settlement was located at 600 m asl. in the humid midlands, in the northwest of the island, in a large area that the ancient local people called Amoco. A place that provided the settlers with land, caves and pastures, as well as the possibility of obtaining water through the use of so-called "horizontal rain", which they collected in small ponds.[2] This according to the *Bimbache* tradition involved the miraculous or saintly tree from which water flowed and which the locals called *Garoé*.[3]

The Norman influence on El Hierro was slight. They were limited for decades to leaving symbolic evidence of their presence, such as the practice of a subsistence economy, as well as developing small extractive activities, on an island, at that time, still not under absolute control by the European settlers. It was not until the incursion of Fernán Peraza the Elder and Captain Juan Machín (1449–1450) that the island was finally pacified, and its true colonization began.

The settlers who arrived on El Hierro from the second half of the fifteenth century onwards proposed a different model to their Norman predecessors. From the simple occupation and extractive strategies of the Normans, they moved to a productive economy. However, in their strategies, the colonists were limited at that time by insurmountable environmental obstacles: the island had no soil and no water, fundamental resources for the implementation of an agro-export economy represented at that time by sugar cane. This condition not only prevented new settlements of colonists, but also incited the desertion of the existing ones, to the point of the Lord prohibiting the departure of those established there to Gran Canaria at a time when this island began to be colonized (Lobo 2019). Hence, this recolo-

nization meant, in its first phase, only modest population growth, because along with the arrival of troops, there was also the departure of many others.

The Castilian reconquest did not mean a break with the indigenous economic model, which was maintained until at least the second half of the sixteenth century with the pre-Hispanic livestock rearing as the dominant productive activity. This provided income for the Lordship in addition to those derived from forestry exploitation (wood and pitch) and the seasonal harvesting of dye plants, such as orchilla and pastel grass, the latter was already being commercialised in the sixteenth century.

And what happened to the indigenous society? Colonization came at a high cultural and human cost. European pathogens, slavery and fighting depleted the local inhabitants' forces. The *Bimbache* chiefs surrendered, and these and some indigenous people took Castilian names and customs. A mostly male colonization meant the unions with local women led to a mestizo and multicultural population. In this way El Hierro, Finisterre and other frontier lands, received old and new Christians, Jews, free men and a few slaves, who coexisted and merged with the free *Bimbache* and who settled in cattle-raising areas. However, in this mixture, indigenous knowledge about the environment survived. Indeed, its use has been decisive for the survival and subsequent food production of the island, as well as its voices, phrases and some native socio-cultural elements that still remain in the documentary and oral memory of the island. At the end of the fourteenth century, El Hierro had a population of eighty neighbours (Bernáldez 1962), around 300 inhabitants. A figure that partially coincides with the forty fathers of a family or 200 inhabitants that Bartolomé García del Castillo would point out centuries later when referring to the population at this time (García del Castillo 2003).

3   The Partition and Its Demographic Influence

The development of sugar cultivation was ecologically impossible, and the island faced a bleak prospect of colonization. Therefore, as an incentive and with the aim of obtaining a few hundred *doblones* from their production, the Lords ceded in the 1500s a part of their private domains: land, water, caves, beehives, etc., on the condition that they were put into production over a period of time, charging little or no rent for their use and only *quintos* (one fifth of the value of produce) and custom duties on the trade of some products: wool, cheese, fish, barley and small livestock. These conditions attracted the population and a small group that benefited from the partition consolidated themselves as an insular ruling class with significant political power and social ascendancy.

---

[2] Horizontal precipitation, mist, hidden or horizontal rain, occurs when the mist passes through the forest canopy pushed by the wind and the water droplets that constitute it are filtered by it, depositing and merging to form larger droplets that end up falling to the ground.

[3] Tree of the lauráceas family, probably a Til or Tilo (Ocotea Foetens), whose leafy branches captured and distilled the water from fog, water that was collected in hollows located at its foot. A storm brought down this legendary tree in 1610. The story of the miraculous tree resonated intensely in the western world for centuries. Tradition has it that it was not the only natural "fog-catcher" on the island.

A dry-stone wall, an *albarrada*, separated land into private and communal uses. Most of the island was initially outside the *albarradas* for common use, but soon (1602) the first usurpations of this communal patrimony took place. In 1637, land outside an *albarrada* (Nisdafe) was divided into two strips in which livestock use alternated with crops during fallow periods, in an attempt to end a historical struggle between farmers and herders over the areas of communal use.

These partitions produced a late repopulation. At the end of the sixteenth century (1585), the island's population increased from 300 inhabitants to 1300 inhabitants (Marco 1943). Its main population centre (Valverde) was established in the aforementioned area known as Amoco, which became the main political and religious centre of the island, as well as the residence of the main beneficiaries of the land distributions. In 1590, Valverde had 250 houses (Torriani 1959) and according to Abreu Galindo at the end of the sixteenth century it had "more than a thousand people" (de Abreu Galindo Fr 1940). This number could well represent practically the whole island's population.

From Valverde, at the end of the sixteenth century, a branch of small settlements around the Nisdafe plateau began to consolidate. To the north and northwest, taking advantage of the supply of water, pasture and firewood, new livestock and agricultural enclaves appeared. To the southwest, the area of San Andrés de Azofa, was a cereal and livestock farming area, with more than a hundred neighbours benefiting from the Açof spring, the pools of water and the nearby pastures. The rest of the island in the sixteenth century was still sparsely inhabited. The south and southwest, with less soil and scarce water, remained as a pasture area with some semi-permanent pastoral enclaves. This is the case of La Dehesa, home since 1546 to the patron saint of the island, the Virgen de los Reyes.

4 Demographic Effects of the Canary Islands Economic Model on the island's Population

The bonanza of agro-export production in the Archipelago during the sixteenth and seventeenth centuries relaunched the economy on other islands. Many areas were turned over to sugar cane and vine cultivation. The commercial fleets that were provisioned in the Canary Islands and the military defences of the archipelago required agricultural and livestock products for their basic needs and El Hierro, together with other islands, provided these. Cereals (barley) and pulses were sent from El Hierro to Tenerife and Gran Canaria once internal demand had been satisfied. As the dominant economic activity, livestock farming was of considerable importance, providing meat, cheese and cured meats, as well as live cattle. Already in the seventeenth century, there was no shortage of *Herreño* wine and brandy on the islands with agro-exporting economies.

In this context, the population increased due to mainland and regional immigration. The Census of the Inquisition of 1605 indicated that Valverde had already reached 250 neighbours, around 1000 inhabitants (Lobo Cabrera 1984–1986). As the population increased, it expanded towards the north, where cattle rearing alternated with agriculture, and with disperse populations developing in the surrounding areas (Rumeu 1947).

The Synods of the Bishop Cámara and Murga in 1629 indicated a population of 600 neighbours, more than 2700 inhabitants (Diaz and Rodríguez 1987). This was significant population growth, a consequence of new arrivals in the early decades of the seventeenth century. At that time, the island was growing at a rate of 1.05% (1590–1680), a figure well above the Archipelago average of 0.73% (Macías Hernández 1988). Bishop Bartolomé García Jiménez's census (1676–1689) confirmed these rising figures, with numbers of around 3500–4000 inhabitants by the end of the seventeenth century, and even reaching 4500 inhabitants in 1680 (Sánchez Herrero 1975).

Díaz Padilla and Rodríguez Yanes (1990) using the parish registers of 1680 showed how Valverde was the main population centre with 939 inhabitants in that century. It was the home of the landowning class, with a significant number of two-storey houses. The area that borders Nisdafe to the north and northwest and which sourced Barlovento or Las Vegas, already showed outstanding production linked to areas of new agricultural land and pastoral activity. The region of Barlovento had at this time more population than Valverde, 1078 people spread in an increasingly disperse way from the town to the border with the Valley of El Golfo. The dwellings in this area were made of stone and tiles, yet there was no population nucleus. The rest of the island was still sparsely populated with 1131 inhabitants divided between San Andrés de Asofa with 758 and El Pinal de S. Antonio (El Pinar), which by this time was reclaiming land from the mountains and had 318 inhabitants. El Golfo was still far from being the prosperous population that it would be in the future. Most of the island was inhabited by a pastoral and transhumant population, with only a few straw houses and cave dwellings, and there was only one house in the whole valley of El Golfo. Another aspect of interest was that most of these dwellings were without doors, testifying to their temporary nature and indicating seasonal residence changes in order to obtain a balance in the production of basic necessities.

5 Economic Crises and Their Impact on the Population of El Hierro

In the mid-late seventeenth century, the population dynamics changed from rapid growth due to immigration to only slight growth or stagnation based on natural population dynamics and a relative emigration effect. This situation coincided with locust plagues (1698, 1703 and 1726); hurricanes and storms (the Virgin of Los Reyes was named patron saint in 1643 after an intense period of these events); as well as with abundant droughts (from 1741, it was decided to move the island's patron saint every four years from her hermitage in La Dehesa to Valverde because of the lack of rain). All these natural setbacks caused a decrease in agricultural and livestock production, food shortages and poverty among the population (Hernández 1983).

In addition to these frequent natural setbacks (the people of El Hierro, the *Herreños* speak of "times of *virados*" to indicate such hardships.

There was also a series of structural factors of a socio-economic nature. From 1640 onwards, a commercial and productive crisis began in the Canary Islands that would intensify when England began to impose protectionist restrictions on the marketing of Canary Island wine, triggering in the following century the ruin of this agro-export and with it the poverty of the Archipelago. In this context, the main islands stopped buying food products from the other islands due to payment difficulties and above all due to the increase in land devoted to subsistence agriculture in these islands. The fall in demand for food products in the inter-island market (cereals, wine and spirits, fruit, livestock and livestock products) was a blow for El Hierro's economy, which was so dependent on the inter-island market in commercial terms.

The reaction of the large landowners to this critical situation was to make use of their local power and expropriate communal areas (forests, pastures and wastelands) in order to increase their income. This added pressure on small peasants and landless tenants, both in working conditions and in the overpayment of rents, which was used to balance the income lost due to the contraction of the agro-export market. However, an intense social conflict broke out, involving the large landowners who benefited from the manorial distributions and the small landowners, in many cases subsistence smallholders, who even with land, progressively became sharecroppers for the former due to the low profitability of their possessions, and of course the sharecroppers and day labourers without land found that their harsh working conditions worsened. There was also the confrontation of the large landowners with the Lord of the island to avoid the payment of taxes and for definitive control of the island. This loss of communal areas also meant the total or, in some cases, partial incorporation of agricultural activity in pastoral areas. For example, in 1700 free grazing was prohibited in El Golfo, leaving this activity limited to marginal areas, such as Pie de Risco, Guinea and Los Llanillos (Díaz Padilla and Rodríguez Yanes 1990).

These natural circumstances and, above all, the disappearance of traditional forms of production due to the privatization of communal land, the harsh social and labour situation and the surplus of labour due to the non-existence of alternatives on the island, led small landowners and day labourers to emigrate to the main Islands or to Latin America. According to Urtusáustegui (1779): "in Tenerife and America, there are swarms of *Herreños* though they would not have left the island unless forced by their needs" (Lorenzo 1983). Viera, years earlier, spoke of "a large number of young men, and even young women, who annually expatriated, either to work on the other islands, especially Tenerife, or to emigrate to America" (de Viera Clavijo 1772–1783).

This migratory process and subsistence crisis, despite high birth rates, led to stagnation or only slight population increases that continued until the beginning of the nineteenth century. In 1706 the Vicar and Commissary of the Holy Office, Juan García Melo (García de Melo 1706) pointed out that Valverde had 240 neighbours (1104 inhabitants), Barlovento 265 neighbours (1219 inhabitants), with a notable number in San Pedro del Mocanal, while Asofa-El Pinar, in which the *pago* de San Andrés stood out, had 259 neighbours (1000 inhabitants).

El Golfo was the only area with a growing population (55 inhabitants). The reason for this increase was, according to the source, "that there were vineyards". This activity became almost a monoculture in association with fruit trees after the disappearance of livestock in communal areas of the Valley. In an attempt to maintain Canary Island wine as a commercial product, from 1700 onwards, a regional commitment was made to produce wine at lower costs, opening up new wine-growing areas. El Golfo in El Hierro fulfilled perfectly this option, and from the eighteenth century onwards the wine-growing activity in the valley consolidated with a corresponding seasonal population, and the chapel of Our Lady of Candelaria was built as a parish church in 1776.

The other alternative to the regional wine decline was to produce spirits, with a higher value than wine, in which El Golfo also participated. Hence, sources (1785) indicate: "the estate of El Golfo, where a lot of wine is produced "*de vidueño*" which they use to produce the best quality brandy [...], which is bought by the merchants of Tenerife and sold to America" (Darias 1943).

In 1719, the island had, according to reports from the Islands' bishops to Rome, 3,080 people (Escribano 1987). Between 1743 and 1744, Bishop Juan Francisco Guillén Isso in his visit to El Hierro noted that the island had 3687 inhabitants, noting that most of the houses were earthen and covered with straw (Guillén Isso 1743–1744). In 1768, the Census of Aranda indicated a population of 4022 (Jiménez de Gregorio 1968). The eighteenth century ended with the Floridablanca Census (1787) that counted 4040 people living on the island (Jiménez de Gregorio 1968).

Since the seventeenth century, practically all the island's settlements had been established, as well as their traditional delimitations or districts: the Villa de Valverde; El Barrio (also called Barlovento or El Norte), Asofa, El Golfo and El Pinar. The statistics of the Marquis of Tabalosos in 1776 (Rumeu de Armas 1943) and Viera y Clavijo in his News (1772–1783) gave quite a complete picture of the settlements. Valverde maintained its insular centrality as an agricultural and administrative centre. The extension of the cultivation and pasture areas to the north consolidated the settlements already existing in this area, expanding with secondary settlements. To the southwest, the intensification of agricultural and livestock farming led to the growth of Asofa. San Antón del Pinal (El Pinar) also increased its agricultural land with fig trees, which helped the main settlements grow. In addition, there was "the very fertile Gulf Valley" (de Viera Clavijo 1772–1783), which saw its settlements grow beyond the traditional ones in the eastern half of the island with others to the centre and west.

During the nineteenth century, the population of El Hierro grew very little. In 1802, the island had 4006 inhabitants (Arbelo 1990), and it did not exceed 5000 inhabitants until the middle of that century (see Fig. 1). The definitive ruin of regional wine production, and the return to the original situation of cultivation to supply the domestic market was a serious setback due to competition among the wines and spirits of the island. This had a notable influence on the population development.

From a social point of view, the tensions between El Hierro's oligarchy and the Lordship came to an end with the disappearance of the latter in 1812. Almost immediately, a hegemonic struggle began between factions of the ruling class, who alternatively exercised dominance over the rest of the population, thanks to the relationships of dependence that bound the peasants to them (administrative favours, etc.). This intensified the pressure on the peasant population and small landowners, with the expropriation and auctioning of more communal land. This created a situation of over-exploitation, which when it coincided with regional crises or periods of drought or plagues made existence extremely arduous, pushing people to emigrate.

Expansions and contractions of El Hierro's economy and emigration explain the population figures in the twentieth century. The censuses of 1900 and 1910 indicate growth rates well below the regional average, synonymous with poverty and emigration. This also coincided with the establishment of a free port regime and the abolition of protective tariffs on local agricultural production, all very negative circumstances for commercial production on the island.

This migratory trend came to a halt in 1910. In this decade, the number of emigrants from the Canary Islands to Cuba exceeded the employment forecasts, resulting in a shortage of workers and social conflict. The Cuban government reacted and only took in seasonal workers. This meant the return of the local population with money to El Hierro, who then invested in land, contributing to a small economic improvement, which helped increase the population to 8,344 inhabitants according to the 1920 census (Fig. 2).

From this decade onwards, sugar prices fell sharply, the Cuban banking collapse took place and in 1929 the well-known world depression occurred. This combination of

Fig. 1 Demographic evolution of El Hierro in the pre-statistical stage. *Source* Itac, Self-elaboration

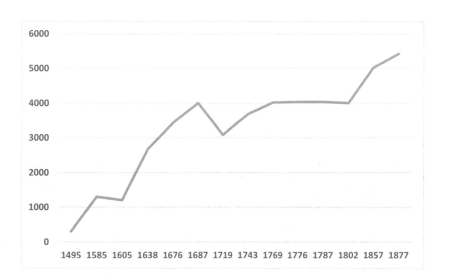

Fig. 2 Map of the evolution of settlement on El Hierro. Self-elaboration

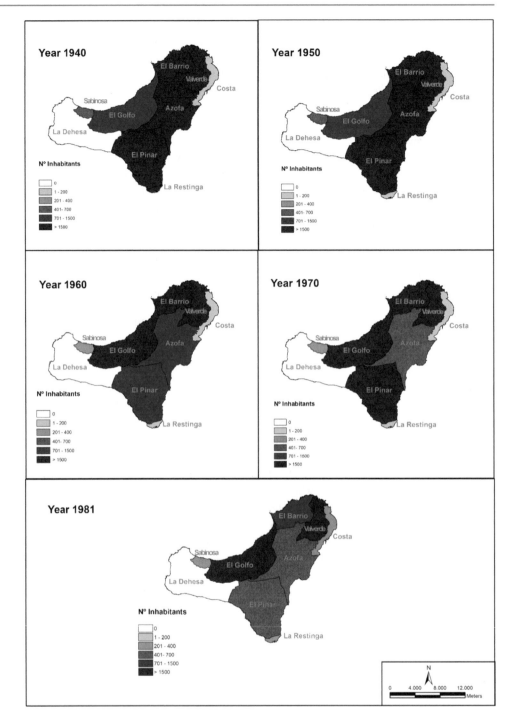

events put an end to Cuba as an ideal destination for any unfortunate *Herreños*. In the Canary Islands, the economic reactivation, after the strong depression caused by the First World War, was halted by the effects of the Great Depression. The result was the return of many of the local population to El Hierro with the money obtained during their work in South America only to find a complex situation there, as the archipelago was also affected by this global crisis.

In the 1930s, the protectionist measures adopted by the main consumer countries of what was then the main export product (bananas) and the decrease in port activity, once again generated a serious setback for the island's economy, which had its sources of financing closely related to the markets of the urban, port and commercial centres of the main islands. This situation was conducive to mass emigration, but this time there was no exodus. The Cuban option was closed and Venezuela, the next preferred destination, was still an incipient agricultural economy with laws against immigration. Without destinations and with substantial domestic agricultural production, the island of El Hierro increased its population extraordinarily, from 8344 inhabitants in 1920 to 9500 and 9810 in 1930 and 1940, respectively. These increases occurred without changes in its socio-productive structure and with only the use of public works as a mechanism to combat unemployment.

6 The Diaspora

El Hierro began the 1940s suffering from a series of setbacks for its development. On the one hand, population growth had generated a volume of workers that could not be absorbed by the island's economic system, even more so when public works in this decade were reduced to the point of their virtual disappearance. In addition, there was political repression, the absence of infrastructures and persistence of traditional forms of farming with low yields, which depended on the overexploitation of labour as the main productive force. Moreover, from a regional point of view, the economic and commercial restructuring that occurred because of the weakening of the free port regime was a severe blow to the urban economies of the main islands: the main consumers of the El Hierro's agricultural products. Nor was the weather favourable during these years, and the island was subjected to one of the greatest droughts of the twentieth century, the so-called "*Seca del 48*". This whole panorama plunged El Hierro into a situation of mere subsistence and its only recourse was emigration. Due to the close historical links, together with the now attractive Venezuelan economic situation, once again many *Herreños* began to think of America as a saviour for their poor situation.

However, emigration was not easy. The Venezuelan demand for labour and the economic needs of the *Herreños* clashed with the difficult circumstances of the post-civil war and the Second World War. The rupture of diplomatic relations between Spain and Venezuela, as well as the de facto ban on foreign emigration by the Franco regime left clandestine emigration as the only option. Fortunately, from 1948, with recognition of Franco's dictatorship by the Venezuelan dictatorship, emigration was legalized, but now there were other drawbacks: obtaining a passport, exit permit, bank deposit and especially the price of the ticket, a fortune for the peasants of El Hierro. Hence, the illegal option remained as the only possible one for many of the island's population.

When, in 1950, it was decreed that emigration was free of charge and the obligation to present economic certifications, letters of call and work contracts were waived, a situation of true diaspora was produced. In three decades (1940–1970), the island lost more than 3000 inhabitants and registered negative migratory balances higher than the provincial average (see Table 1). Young men at first, and from 1950 onwards with the Regrouping Plan, entire families left for Venezuela. As a result, El Hierro became an ageing island with many empty houses.

7 Changing Trends

However, from the 1970s onwards, there was a new demographic trend because of three fundamental factors: the development and use of irrigated agriculture in the Valley of El Golfo, the flow of state capital into public works and the progressive tertiarization of the population. These changes produced a transformation in El Hierro's labour structure, from the dominance of sharecropping relationships to mostly salaried forms of labour, either in irrigated agriculture or the service sector. In this way, as the economic situation changes, the traditional layout of human settlements, reflecting a socio-economic organisation that was becoming obsolete, was broken down.

The midland zones of the island, historically agricultural and livestock farming areas and the location of its main centres, underwent a profound change. The populations located there have declined and aged, while the Valley of El Golfo has gained demographic importance, as it increasingly becomes a centre of prosperous agricultural exploitations. This economically more active area may replace Valverde in importance, the latter retaining some relevance as an

Table 1 Distribution of the population by county (1940–1970)

	1940	1950		1960		1970	
Valverde	1854	1687	− 167	1636	− 51	1443	− 193
El Barrio	2041	1659	− 382	1591	− 68	1078	− 513
Asofa	1625	1542	− 83	1374	− 168	669	− 705
The Gulf	1727	1673	− 54	1981	308	1363	− 618
El Pinar	1602	1621	19	1375	− 246	950	− 425
Total	8849	8182	− 667	7957	− 225	5503	− 2454

Source Nomenclator, INE. Self-elaboration

Table 2 Distribution of the population by districts (1981–2011)

	1981	1991		2001		2011	
Valverde	1841	2007	166	2447	440	2796	349
El Barrio	963	978	15	1256	278	1493	237
Asofa	670	541	− 129	629	88	759	130
The Gulf	1940	2408	468	3436	1028	4143	707
El Pinar	994	1061	67	1655	594	1804	149
Total	6408	6995	587	9423	2428	10,995	1572

Source Nomenclator, INE. Self-elaboration

Fig. 3 Panoramic view of El Golfo

administrative centre. From the end of the 1980s, until the segregation of El Pinar as a municipality in 2007, Frontera became the most populated municipality on the island (Table 2).

Another characteristic of the new population situation has been the extraordinary growth of historically unpopulated coastal areas, which since the mid-1970s have become vital points for the service economy (La Restinga) or for second homes (the coast between El Tamaduste and Timijiraque, in Valverde) (Fig. 3).

Overall, the main contemporary demographic event on the island has been the intensification of the immigration process due to the difficult and convulsive economic and political situation in Venezuela. The return of people who once left the island, together with second or third generations of those who left, to which must be added a relative increase in resident Europeans (Germans and Italians) are responsible for the continuous population increase in the last third of the

twentieth century and the first decades of the twenty-first century, which means El Hierro's population now stands at more than 11,000 inhabitants.

References

Arbelo Curbelo A (1990) Población de Canarias, siglos XV al XX y sus fenómenos demográficos sanitarios 1901–1981. Imprenta Pérez Galdós

Bernáldez A (1962) Memorias del reinado de los Reyes Católicos. Blass, S.A. Tipográfica

Darias y Padrón DV (1943) Los Condes de La Gomera (documentos y notas históricas). Revista de Historia 16(64):300–315

de Abreu Galindo Fr J (1940) Historia de la conquista de las siete islas de Gran Canaria. Imprenta Valentín Sanz

de Viera Clavijo J (1772–1783) Noticias de la Historia General de las Islas Canarias. Imprenta de Blas Román

Díaz Padilla G, Rodríguez Yanes JM (1990) El Señorío en las Canarias Occidentales. La Gomera y El Hierro hasta 1700. Excmo. Cabildo Insular de El Hierro y Excmo. Cabildo Insular de La Gomera

Escribano Garrido J (1987) Los jesuitas y Canarias, 1566–1767. Facultad de Teología Universidad de Granada. ISBN 84-85653-51-3

García de Melo J (1706) Memoria de los lugares y vecindad que se componen en la isla de El Hierro. Archivo de Acialcázar. We thank Dr. Germán Santana Pérez for the reference

García del Castillo B (2003) Antiguas Ordenanzas de la isla de El Hierro. Cabildo Insular de El Hierro. ISBN: 84-920983-9-2

Guillén Isso JF (1743–1744) Extracto de la visita del obispado de Canarias por el Ilustrísimo Juan Francisco Guillen Isso. Archivo de Acialcázar. We thank Dr. Germán Santana Pérez for the reference

Hernández Rodríguez G (1983) Estadística de las Islas Canarias 1793–1806 de Francisco Escolar y Serrano. Caja Insular de Ahorros. ISBN: 84-7580-004-1

Jiménez de Gregorio F (1968) La población de las Islas Canarias en la segunda mitad del siglo XVIII. Anuario De Estudios Atlánticos 1–14:127–301

Jiménez Gómez MC (1993) El Hierro y Los Bimbaches. Centro de la Cultura Popular Canaria. ISBN: 84-7926-090-4

Junyent C (2013) Entre lajiales y brumas. Cabildo de El Hierro – Ciència en societat. ISBN: 978-84-616-3576-4

Le Canarien (1959) Crónicas francesas de la Conquista de Canarias. (Fontes Rerum Canariarum, VIII). Instituto de Estudios Canarios. El Museo Canario

Lobo Cabrera M (1984–1986) El Tribunal de la Inquisición de Canarias intento de traslado a Tenerife. Revista de Historia Canaria 174:107–114

Lobo Cabrera M (2019) El Hierro. De la Prehistoria a la Colonización. Mercurio Editorial. ISBN: 978-84-17890-55-1

Lorenzo Perera MJ (1983) Diario de viaje a la isla de El Hierro en 1779. Centro de Estudios Africanos. ISBN: 8460031330

Macías Hernández AM (1992) Expansión europea y demografía aborigen: El ejemplo de Canarias 1400–1505. Boletín de la Asociación Española de Demografía Histórica X(2):9–45

Macías Hernández AM (1988) Fuentes y principales problemas metodológicos de la Demografía Histórica de Canarias. Anuario De Estudios Atlánticos 34:51–158

Marco Dorta E (1943) Descripción de las Islas Canarias hecha en virtud de mandato de S.M. por un tío del Licenciado Valcárcel. Revista De Historia 63:197–204

Morales Padrón F (1978) Canarias crónica de su conquista. Excmo. Ayuntamiento de Las Palmas de Gran Canaria y Museo Canario. ISBN: 8450029511

Rumeu de Armas A (1943) Una curiosa estadística canaria del siglo XVIII. Rev Int Sociol 4:179–185

Sánchez Herrero J (1975) La población de las Islas Canarias en la segunda mitad del siglo XVII (1676 a 1688). Anuario De Estudios Atlánticos 21:237–415

Torriani L [1590] (1959) Descripción de las Islas Canarias. Goya Ediciones

Rural Landscapes in an Oceanic Volcanic Island

Víctor Onésimo Martín Martín

Abstract

For a quarter of a century, the territorial dynamics of El Hierro have been influenced by the tertiarization of its economy, fundamentally focused on non-intensive tourist activity (green or nature tourism, hiking, rural tourism, sports in nature), causing a gradual abandonment of agricultural activities and, therefore, of their agricultural landscapes. However, it is still possible to recognize today the traces of traditional agricultural landscapes of great interest, both for heritage and for food production. It is then necessary to carry out a typology of these agro-cultural spaces with the aim of delimiting them, knowing their production systems, conserving and organizing them and, what is more remarkable, proposing for the future those ways of managing them. That can contribute to the sustainable development of the island, further justifying the declaration of the Island as a Biosphere Reserve.

Keywords

Agro-pastoral-forest landscape • Agricultural abandonment • Tangible and intangible heritage • Oral history • El Hierro

1 Introduction

Initially, when the first humans arrived on the island of El Hierro, they found an environment that would condition the use of its natural resources. The pre-colonial occupation by the *Bimbapes* (indigenous people) and the subsequent colonial occupation with its two stages (modern or feudal

lordship and contemporary or bureaucratic capitalism) had to rely on these natural conditions to ensure survival as a society. The agro-pastoral-forestal use of El Hierro's territory has given rise to unique agrarian landscapes, some born early on and others arising barely half a century ago.

For the last quarter of a century, the territorial dynamics of the island has been influenced by the tertiarisation of its economy, fundamentally focused on non-intensive tourism (green or nature tourism, hiking, rural tourism, nature sports), causing a gradual abandonment of agricultural activities (see Table 1) and, therefore, of its agrarian landscapes.

However, it is still possible to recognize today the traces of some traditional agricultural landscapes of great interest both in terms of heritage and food production for the achievement of the so-called food sovereignty of the island, in particular, and the Canary Islands in general.

It is therefore necessary to carry out a typology of these agro-cultural spaces with the aim of delimiting them, understanding their production systems, conserving and ordering them and, more importantly, proposing ways of managing them in the future that can contribute to the sustainable development of the island, further justifying the declaration of the island as a Biosphere Reserve. This is all the more pressing in that the abandonment of the forests and the natural death of the people who manage them are causing the disappearance of a vernacular knowledge of unquestionable intangible heritage value.

2 The Role of Ecological Determinants in the Configuration of the Agrarian Landscapes of El Hierro

The genesis and evolution of the traditional agricultural systems of El Hierro involved decisive human factors—its history—that had to overcome a series of ecological or environmental conditioning factors that, far from favouring agricultural activity, forced El Hierro and the *Herreña*

V. O. Martín Martín (✉)
Research Group on Underdevelopment and Social Backwardness GISAS-Department of Geography and History, University of La Laguna, San Cristóbal de La Laguna, Spain
e-mail: vbmartin@ull.es

J. Dóniz-Páez and N. M. Pérez (eds.), *El Hierro Island Global Geopark*, Geoheritage, Geoparks and Geotourism, https://doi.org/10.1007/978-3-031-07289-5_7

Table 1 Area of current crops, pastures and abandoned crops on the island of El Hierro by municipality (2015)

	El Pinar	Frontera	Valverde	El Hierro
Higuera	182.3	0.65	7.93	190.88
Almond tres	12.3	0.09	0.74	13.13
Other temperate fruit trees	23.61	12.48	37.81	73.9
Vineyards	51.49	103.89	47.47	202.85
Banana plantations	12.17	49.03	0.06	61.26
Family garden	6.69	10.91	17.23	34.83
Pope	5.15	5.14	22.04	32.33
Clean orchards	4.47	10.78	15.32	30.57
Tropical pineapple	0.07	116.63	0.05	116.75
Avocado	0.95	4.13	5.33	10.41
Mango	0.19	24.44	2.75	27.38
Other subtropical fruit trees	0.64	15.71	7.2	23.55
Cereals and corn	1.03	0.92	62.73	64.68
Vegetables	0.38	3.27	1.71	5.36
Citrus fruit	0.13	4.48	0.76	5.37
Total cultivated agricultural area	301.57	362.55	229.13	893.25
Pasture	18.8	470.4	1588.94	2078.14
Tagasaste	1.36	0	118.98	120.34
Total pastures	20.16	470.4	1707.92	2198.48
Total abandoned agricultural area	818.95	441.22	1688.3	2948.97

Source El Hierro Crop map 2015 https://www.gobiernodecanarias.org/agriculture/themes/crop_map/el_hierro/el_hierro_2015.html. Note how 71% of the agricultural area, excluding pastures, is abandoned

society to develop complex agro-pastoral-forestry systems that have given rise to a great diversity of agricultural landscapes of undoubted heritage value. The following is a summary of these conditioning factors: (a) topographical, (b) geological, (c) geomorphological, (d) edaphological, (e) climatic and (f) hydrological.

(a) Topography

El Hierro's small surface area is compounded by its high average altitude, which results in significant slopes, as well as the small extension of its many coastal areas (between 0 and 300 m above sea level). These are limited to the El Golfo and Las Playas landslide valleys and the small "low island" of Hoya de El Verodal, the generalised cliffs of its coasts and the scarce development of a network of ravines. El Hierro is shaped volumetrically as a truncated triangular pyramid with three concave lateral faces, Julan-El Golfo-Las Playas, which can reach altitudes of more than 1000 m, and a high upper base (above 600 m), formed by two plateau areas, La Dehesa and Nisdafe, sloping progressively towards the SW and NE, respectively.

The unique topography of the island presents conditioning elements for agricultural use: (1) difficulty for subtropical crops due to the absence of coastal plains, (2) almost vertical cliff walls impossible to cultivate (hence, compared

to the other islands that also have significant altitudes, El Hierro does not have large areas of dry stone terraces) and (3) the predominance of the so-called dry farming of "medianías" (midlands) and high altitude pastures (both plateaus).

(b) Geology

It should be taken into account that during the Quaternary and in relatively recent times a very intense volcanic activity of a mainly basaltic nature has affected the island, practically in its entirety. Three volcanic cycles have taken place with hardly any periods of eruptive calm, the last two being very recent (intermediate and recent series), so the volcanic shapes (cones with their craters and calderas) and eruptive materials (fields of basaltic lavas of malpaíses and "lajiales" and of lapilli -called `jables' on El Hierro-) are preserved with few alterations. However, such young geological materials have represented serious limiting factors for agricultural and livestock uses, as they have been scarcely altered by vegetation and physical and chemical processes.

(c) Geomorphology

The predominance of recent morphostructures as opposed to volcanic morphosculptures explains the lesser importance of erosive phenomena, exemplified by the low level of the ravine network on the island. The exceptions are the macroforms of the three gravitational landslide valleys (Julan, Las Playas and El Golfo). In them, the dynamics of the slopes have caused the deposition of sediments (sands, gravels and pebbles) in the foothills of their almost vertical escarpments. In the absence of soils on the island, these sediments have historically been used for agricultural activities (terracing) and livestock (pastures).

(d) Soil science

Following the work of Fernández in this section (Fernández et al. 1974), the recent nature of the volcanic materials that cover the island means that they are little altered and the soils, therefore, are little developed, as they correspond to the first phases of the alteration of the volcanic soils. The most important edaphic formations are represented almost exclusively by andosols (vitric), located above 500 m. Below this isohyet, there are poorer, carbonate soils of a pulvurulent nature. On the island as a whole, there are no soils with textural horizons. However, in some areas near Valverde, around 700 m above sea level where, exceptionally, accumulations of clays appear, due to endorheism phenomena caused by the interposition on the slope of lava flows or volcanic edifices.

The lithosols are very abundant and extensive, formed by "malpaíses" (recent, slightly altered lava flows) and recent cones and fields of lapilli. The erosion lithosols appear on the slopes of very rugged topography. Thus, in the El Golfo valley, there are large colluvial surfaces, sandy-stony, poorly evolved, forming cones and dejection fans that can be described as entisuelos.

In the cornice of El Golfo, with a steep slope, the laurel forest reaches up to 800 m, with thin and poorly evolved ranker-arid erosion soils. However, lack of soils has inspired the ingenuity of the *Herreño* farmers to try to overcome it, giving rise to original agro-ecosystems.

(e) Climate

Although El Hierro shares the same climatic characteristics as Canary islands of greater relief, its terrain leads to some differences that influence its agricultural use. Due to its shape of a truncated pyramid, whose upper base extends between 600 and just over 1000 m above sea level (almost two thirds of the island's surface), the "*medianías*" (mid-lands) are the greatest surface area among the three altitudinal bioclimatic floors (coasts, "*medianías*" and summits). This extensive area of "*medianías*" is in the condensation area of the trade winds, whose humidity also affects the midlands of the south of the island due to overflow because

of the reduced surface area above 1400 m. In addition, the high leeward plain of Taibique-Las Casas-Julan East is favoured by its opening towards the humid winds from both NE and NW.

It is no coincidence that, given the stability of the prevailing socio-economic conditions for five hundred years, the traditional settlements of El Hierro are located in these mid-altitude and high-altitude lands where, given the absence of irrigation water, the constant humidity of the trade winds has allowed agriculture and pastures in an otherwise dry regime.

(f) Hydrology

The surface water resources of El Hierro are scarce due to the recent conformation of the island, with a predominance of porous surface substrates and the scarce presence of impermeable materials in the subsoil. For all these reasons, El Hierro has historically been the only island in the Canary Islands with a totally dry farming system. However, every year, the island's aquifers incorporate on average about 11 hm^3 of total rainfall, which shows that groundwater is important. Although only recently (since the late sixties of last century) has the population of El Hierro had access to them through the construction of wells and galleries. The consumption of groundwater is reduced to 1.9 hm^3 with the new irrigated agriculture: the main water consumer (1.49 hm^3) (Felipe and Herrera 2019). Technological advances, belatedly arrived on the island of El Hierro, allowed the creation of the last and most modern insular agricultural landscape.

3 Five Insular Territorial Keys to Understanding the Uniqueness of the Agrarian Landscapes of El Hierro

In the evolution of the construction of El Hierro's agricultural landscapes, there are a number of elements that make this island unique with respect to the rest of the Canary Islands. One of these elements, "*la mudada*' (the move), is an agricultural practice of transhumance that is no longer in use and is difficult to observe if you do not know the history of the island. Three other unique elements can be seen with the naked eye in all the ecological environments of the island: the stone walls, the fig trees and the total agricultural use of the territory. This last feature has to do with the scarcity of water: rainfall and humidity.

The agricultural "*mudada*', together with the ploughing of Nisdafe, responds to the particular form of economic organisation established by the *Herreña* seigniorial class to take advantage of the natural resources and produce both income and the food necessary for the inhabitants of El Hierro (Galván 1997). During the feudal or seigniorial period, the rents for the large landowners (known as *rabos blancos* or white tails) were obtained through the social relationship of the "*medias perpetuas*" (perpetual rights) and a large communal territory of free grazing (*rabos negros* or black tails). The transition from the manorial system to the contemporary system during the nineteenth century did not change the class character of Herreña society, since the "*mudada*" continued to be the form of land use until its decline in the last quarter of the last century. Until the seventies of the twentieth century, important landowners continued to accumulate land through the usurpation and/or purchase of communal land and changed the system of "*medias perpetuas*" to the "*medianería*" (midlands). The shepherd and the poor farmer of El Hierro thus saw the rent they had to pay to large landowners increase, while their free access to communal lands decreased (Franco's colonization policy in the Dehesa Comuna was the last expropriation of that period). In the absence of agrarian reform, emigration was the natural solution to the increasing overexploitation and oppression of the island's poor families.

The "*mudada*" consisted of peasant families moving to the coastal areas at certain times of the year with their livestock, tools and household goods. The most important "*mudadas*" took place from the higher parts of the island to the north (El Golfo), the agricultural "*mudada*", and to the south (Las Playas, Timijiraque and Cardones and the Dehesa Comunal) the pastoral "*mudada*", which was probably the

oldest (Sánchez 2018). The move took place in the winter season, mainly in the month of January, when there was a significant amount of grass for the cattle (coastal pastures), and at the same time, to carry out the agricultural work of digging the vines and pruning in El Golfo. Farmer labourers remained there until the end of February, after which they returned to the highlands to continue the agricultural campaign and to take advantage of the pastures of the high midlands and summits. A second move to El Golfo was repeated during the summer, in August and September, also related to agricultural work and pastures for livestock ("*rastrojeras*" or stubble clearing) and fishing activities such as shellfishing and coastal fishing. The agricultural tasks were mainly the grape harvests (and cereal harvest), which were carried out from the second half of August until the end of September, after which farm labourers returned to the highlands.

Basalt stone walls are distributed throughout the island. They represent the result of the struggle that livestock rearing of a communal nature (free grazing), initially predominant, was losing against agrarian activities of a feudal and semi-feudal private nature (Lorenzo 2011). In reality, the creation of enclosures (plots of land surrounded by stone walls) and cattle tracks is the way in which *Herreño* social groups prevented the introduction of livestock into cultivated areas and stopped their cattle from leaving their properties (Fig. 1).

If you had to highlight a fruit tree of El Hierro, this would undoubtedly be the fig tree. In fact, there are too many fig

Fig. 1 The basalt stone walls of the enclosures and cattle road are a characteristic element throughout the island of El Hierro (Nisdafe)

trees for an island with so few inhabitants, so the fig, especially in the past, has always been one of its main export items. In addition, its leaf serves as green fodder for cattle in summer. The enormous ecological adaptability of this dry land tree, both from the climatic and edaphic point of view, means that we find it, isolated in a dispersed manner, in association with other crops, or in true monoculture plantations, throughout the length and breadth of the island. The fig tree appears from sea level to the summits, on the windward and leeward sides, on the east and west sides, on evolved soils and on recent badlands, in "polvillos" (mixture of soils and lapilli) and on slag and lapilli fields. A "gorona" is a circular wall built to protect fruit trees from livestock and almost always contains a fig tree (Fig. 2).

The limited natural resources of El Hierro and the backward production techniques and societal relations meant that all areas of the island territory have been used by humans over the centuries. Even the island's forests were used by its inhabitants for forestry and livestock exploitation. Therefore, it can be said that there are no natural areas untouched by humans, because the inaccessible in El Hierro does not exist when it comes to subsisting in an inhospitable environment. Even today, the apparent abandonment of agropastoral systems, is just that, appearance, because on an island with more livestock than agricultural must always have a source of food such as grass for livestock.

This element of livestock preponderance since *Bimbape* times, origin of "*la mudada*" is the reason for our final description of the El Hierro's singularity. It is the seasonal provision of natural pastures for livestock according to altitude, and how the livestock moves throughout the territory looking for fresh grass that the high altitudinal gradient (0–1500 m) provides. It is the altitude and the climate that are responsible for the availability of pastures throughout the year. From late autumn to early spring, it is the rain that guarantees the coastal pastures, but from spring to autumn it is the humidity of the trade wind mists (and the horizontal rain) that descends from high altitude that governs the movement of livestock from the forest peaks to the lower northeast-facing mid-altitude lands. Even today, despite the crisis of traditional agro-ecosystems, there are still livestock farmers who drive their herds down the slopes following the mist generated in the area of thermal inversion of the trade winds. In addition, there is the "*juelgo*" or "*manchón*" (mixed cultivation of cereals and leguminous plants—lupins, peas, broad beans—for green fodder), haymaking and *tagasaste* (*Chamaecytisus proliferus*), thus providing essential feed for the cattle in anticipation of bad weather years that can affect the natural reproduction of the pastures.

More than five centuries of colonialism have not managed to convert the El Hierro's shepherds into sedentary farmers, nor come to that, into state or tourism employees.

Although there are few shepherds left, if you dig a little into the mentality of the farmer, the civil servant or the service worker, we still find in these people the roots of the El Hierro's identity: the freedom that has since *bimbape* times meant the availability of free grazing of their livestock throughout the length and breadth of an island that will

Fig. 2 Fig tree cultivation in the high midlands of El Pinar

always want to remain a large communal pasture demo-cratically managed by the locals, themselves.

4 The Agro-Pastoral-Forest Landscapes of El Hierro

The sources consulted and the fieldwork conducted provide an initial approach to the agricultural landscapes of the island, showing their diversity and richness. Many of these landscapes date back to the seventies of the last century. For this reason, we have called them traditional agricultural landscapes. However, it is necessary to start from the typi-fication and delimitation of these traditional agricultural systems because they represent part of the cultural heritage and identity of the island and an element of present and future applications for sustainable production in the primary sector.

For this classification we have differentiated between agrarian landscapes (still recognizable today and of some extension), agrarian landscape enclaves (recognizable, but reduced dimensions) and agrarian paleo-landscapes (practi-cally or totally abandoned). We briefly review the list of the most significant present-day agricultural landscapes of El Hierro and refer the reader to the detailed descriptive table (see Table 2) (Fig. 3).

Within the class of *agricultural landscapes on evolved soils of the "medianías" (midlands)*, these three appear:

1. Rainfed polyculture in the Northeast (Valverde). This is probably the first agricultural landscape after the con-quest, as this is where the best soils on the island and the best climatic conditions for dry farming are to be found. Today, this landscape is in clear regression.
2. Scattered chestnut groves in the high midlands of the El Golfo valley (Frontera). This unique landscape of the

Table 2 Inventory of agricultural landscapes, agricultural paleo-landscapes and landscape enclaves of El Hierro

Category	Class
1. Agricultural landscape on evolved soils of the midlands Humid windward "medianías" (midlands) (500–1000 m altitude) Semi-arid leeward midlands (600–1200 m altitude)	1. Rainfed polyculture in the "medianías" (midlands) of the northeast of El Hierro (Los Barrios, Valverde): cereals (mainly barley; wheat, rye), fruit trees, potatoes, sweet potatoes, millet/corn, legumes and fodder legumes (chickpeas, broad beans, beans, kidney beans,, lentils, peas, peas, lupins, lupins) and vines on evolved soils in the humid windward midlands
	2. Scattered chestnut groves on poorly developed soils in the high "medianías" (midlands) (800–1100 m altitude) of the valley of El Golfo (Frontera)
	3. Rainfed polyculture with tuneras of the Taibique-Las Casas plain (El Pinar)—occasional irrigated land—(cereals, fruit trees, potatoes, vines, forage crops) on poorly developed soils of the semi-arid leeward "medianías" (midlands) of El Hierro
2. Agricultural landscape in lithosols Crops in areas of recent volcanism and little altered materials Crops in areas of recent volcanic materials mixed with poorly evolved soils: "polvillos"	4. Rainfed vineyards in the valley of El Golfo (Frontera), in low "medianías" (midlands) on basaltic scoria ("breñas") (need for "despedregamiento") with scattered stone fruit trees (peaches, apricots, loquats, plums)
	5. Rainfed vineyards of Sabinosa and Frontera (Frontera) on basaltic volcanic cones of lapilli, inclined, without construction of terraces, in low "medianías" (midlands)
	6. Vineyards of Echedo (Valverde) (together with scattered fruit trees: mulberry, fig and pear trees), rainfed on recent lapilli fields with stone walls
	7. Dry-farmed fig tree crops in "polvillos" (on recent, scarcely edaphized basaltic lapilli) of the high midlands of Taibique-Las Casas/Julan Oriental (El Pinar), southwest of the leeward slope
	8. Rainfed vines in "polvillos" (on recent basaltic lapilli, scarcely edaphized) of the Taibique-Las Casas (El Pinar) "medianías" (midlands), southwest of the leeward slope
3. Terraced agricultural landscape on sedimentary slope deposits in ravines of ancient volcanic massifs	9. Polyculture in rainfed terraces (potatoes, rye, beans, broad beans, chickpeas, millet/corn, pumpkins, juelgo/manchón) in the low "medianías" (midlands) of El Golfo in Frontera and Sabinosa (Frontera) on sedimentary deposits of slope
4. Agricultural landscape of subtropical fruit trees	10. Open-air tropical pineapple and banana crops in greenhouses (and other tropical fruit trees, to a lesser extent) under irrigation on recent former "malpaíses" (aa lava flows) in the coastal platform of the El Golfo valley (Frontera) with "suelos de prestación", transported from the plateau of Nisdafe

(continued)

Table 2 (continued)

Category	Class
5. Livestock landscape	11. Livestock landscape of the Dehesa Comunal (Frontera, El Pinar): sheep (predominant), goats and cattle (to a lesser extent) in flat arid pastures and high leeward areas (variant: distribution of El Cres in Frontera)
	12. Livestock farming landscape of the Nisdafe plateau (Valverde): sheep (predominantly), cattle and goats (to a lesser extent) in pastures in flat and high windward wetlands; intercropping (barley, leguminous plants, winter potatoes and `papas de hoyo' or *veraneras*; `juelgos')
	13. Landscape of the tagasaste bush pasture plantations of the Isora "*lomadas*" (flat surfaces between ravines) (Valverde), in the high leeward "*medianías*" (midlands)
	14. Livestock-forest landscape of the Fayal-Brezal green woodland (Frontera, El Pinar)
	15. Livestock-forest landscape of the summit pine forests (El Pinar)
Paleo-landscapes and agrarian landscape enclaves	Paleo-landscapes • Agro-pastoral paleopastoral landscape with "*goronas*" or "*góranes*" with fig trees on recent "malpaíses" (volcanic flows) in the coastal of El Golfo (Frontera) • Paleo-landscape of "*henequén*" (sisal) cultivation on poor soils on the North Coast (Pozo de las Calcosas) • Paleo-landscape of cereal-grassland crops on lithosols and poor soils of the North and Northeast coast (Valverde) • Paleo-landscape of natural coastal grasslands on sedimentary slope deposits in the gravitational landslide valley of Las Playas (Valverde) • Paleo-landscape of the landscape enclave of intensive agriculture of irrigated banana trees on "*malpaíses*" with "*suelos de prestación*", in the low island of Punta del Verodal (Frontera) Landscape enclaves • Landscape setting of rainfed almond tree crops in "*polvillos*" (on recent, scarcely edaphized basaltic lapillis) in the southwest of the high "*medianías*" (midlands) of the leeward slope of Taibique-Las Casas (El Pinar) • Landscape enclave of banana plantations in greenhouses on "malpaíses" (volcanic flows) with "suelos de prestación" on the low island of Tacorón (El Pinar) • Landscape enclave of fig trees with prickly pear cactus on recent "*malpaíses*" (volcanic flows) in Los Llanillos (Frontera) • Landscape enclave of tropical fruit trees (mango, papaya, avocado) on slope deposits, in Las Lapas in the valley of El Golfo (Frontera)

Source Prepared by authors based on documentary, cartographic, statistics, oral information and field work

high midlands of the most mountainous islands is represented in El Hierro in the valley of El Golfo, but much more restricted due to the absence of quality soils. Nowadays, it is difficult to observe, as it is in an advanced state of abandonment within the fayal-brezal.

3. Rainfed polyculture with prickly pear cactus in the Taibique-Las Casas plain (El Pinar). This is a typical landscape of the high leeward midlands. Here, it stands out for the large presence of prickly pear cactus (used as fodder for livestock and for its fruits both fresh and dried locally known as "*porretas*"), sown on the edge of plots, which gives them a certain uniqueness among the Canary Island agrarian landscapes.

In the class *of agricultural landscapes in lithosols,* we highligh those with crops in areas of recent volcanism and little altered materials:

4. Vineyards with stone fruit trees (peaches, apricots, loquats, plums) on dry land on terraced slag in the midlands of the El Golfo valley (Frontera). This agricultural landscape, which began to be built from the expansion of the vineyards in the sixteenth century, has always been linked to the worst soils (volcanic substrates), fruit trees often accompany the vines. It can be said that the cultivation of vines, although it has diminished in this region, currently represents an active agricultural landscape (Fig. 4).

5. Monoculture of unirrigated vineyards of Sabinosa and Frontera on basaltic volcanic cones of lapilli little altered and not terraced (Frontera). On these cones of lapilli and slag, the vines appear as a monoculture without fruit trees to accompany them and without the construction of terraces. There are some abandoned plots, but the landscape is still cultivated.

Fig. 3 Map of the agrarian landscapes of El Hierro. *Source* Prepared by authors based on the sources consulted: documentary, cartographic, statistics, oral information and field work

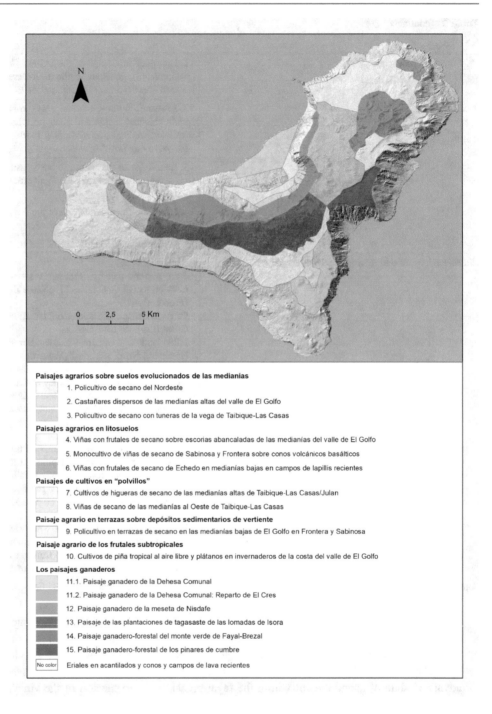

Paisajes agrarios sobre suelos evolucionados de las medianías

1. Policultivo de secano del Nordeste

2. Castañares dispersos de las medianías altas del valle de El Golfo

3. Policultivo de secano con tuneras de la vega de Taibique-Las Casas

Paisajes agrarios en litosuelos

4. Viñas con frutales de secano sobre escorias abancaladas de las medianías del valle de El Golfo

5. Monocultivo de viñas de secano de Sabinosa y Frontera sobre conos volcánicos basálticos

6. Viñas con frutales de secano de Echedo en medianías bajas en campos de lapillis recientes

Paisajes de cultivos en "polvillos"

7. Cultivos de higueras de secano de las medianías altas de Taibique-Las Casas/Julan

8. Viñas de secano de las medianías al Oeste de Taibique-Las Casas

Paisaje agrario en terrazas sobre depósitos sedimentarios de vertiente

9. Policultivo en terrazas de secano en las medianías bajas de El Golfo en Frontera y Sabinosa

Paisaje agrario de los frutales subtropicales

10. Cultivos de piña tropical al aire libre y plátanos en invernaderos de la costa del valle de El Golfo

Los paisajes ganaderos

11.1. Paisaje ganadero de la Dehesa Comunal

11.2. Paisaje ganadero de la Dehesa Comunal: Reparto de El Cres

12. Paisaje ganadero de la meseta de Nisdafe

13. Paisaje de las plantaciones de tagasaste de las lomadas de Isora

14. Paisaje ganadero-forestal del monte verde de Fayal-Brezal

15. Paisaje ganadero-forestal de los pinares de cumbre

No color | Eriales en acantilados y conos y campos de lava recientes

6. Vineyards with fruit trees (mulberry and fig trees) in the dry land of Echedo in the lower midlands on recent lapilli fields with stone walls (Valverde). These scattered vineyards cultivated in deep fields of lapilli and slag stand out for the black colour of their substratum (as opposed to the ochre ones, the result of oxidation and greater age of the previous landscape) due to the recent volcanic materials emitted (reminiscent to a certain extent of the landscapes of the vineyards of Lanzarote). After decades of abandonment, nowadays there are some farms that have been recovered for this crop.

Secondly, cultivated landscapes in areas of recent volcanic materials mixed with poorly evolved soils ("polvillos"):

7. Cultivation of rainfed fig trees in "*polvillos*" in the high midlands of Taibique-Las Casas/Julan Oriental (El Pinar). This landscape is unique in the Canary Islands, as the fig trees occupy the whole plot as a monoculture in rows and with a plantation frame (possibly some cereal crops in the past). Their origin seems to be a distribution of wasteland following the disentailments of the nineteenth century in

Fig. 4 Agricultural landscape on recent basaltic slag in the *"medianías"* (midlands) of the El Golfo valley. Note the nearby Monteverde vegetation (fayal-brezal) on which the arable soil was built

the context of the boom in exports of figs in the past (Hernández and Niebla 1985). At present, this fruit landscape is abandonned; nevertheless, it is possible to recognize its former remarkable extension.

8. Rainfed vineyards in *"polvillos"* on recent basaltic lapilli, scarcely edaphized, of the *"medianías"* to the West of Taibique-Las Casas (El Pinar). The vineyards are scattered among others formerly dedicated to polyculture, and although the latter are in a considerable state of abandonment, the cultivation of vines has survived and has even been strengthened and modernized in recent years.

There is also an example of a kind of agricultural landscape on terraces on sedimentary deposits on slopes in ravines of ancient volcanic massifs:

9. Polyculture on dry terraces in the lower midlands of El Golfo in Frontera and Sabinosa on sedimentary slope deposits (Frontera). This is a landscape developed on sands, gravels and pebbles formed after the gravitational collapse that gave rise to the El Golfo valley. Given the origin of the sediments, it is necessary to clear and terrace the land to organize the crops (although we do not observe here the large, terraced slopes of other islands such as La Gomera).

A final example of a landscape related to agricultural activity is that of the *subtropical fruit trees agrarian landscape* class:

10. Open-air tropical pineapple and banana crops in irrigated greenhouses on the coastal platform of the El Golfo valley (Frontera). This is the most recent agricultural landscape on El Hierro. Its creation has been possible thanks to the availability of water for irrigation from the seventies of the last century and the *"suelos de prestación"* (transported soil) from the plateau of Nisdafe. Both elements have transformed these recent former *"malpaíses"* (aa lava flows) into a landscape of intensive agriculture whose production of bananas and tropical pineapples (and a few hectares of other tropical fruits) are mainly exported (Fig. 5).

Finally, as it could not be otherwise on an island with a long pastoral tradition (mainly sheep), we find up to five landscapes dominated by the livestock component (livestock-agricultural, or livestock-forest). These are the classes of *livestock landscapes*:

11. Livestock landscape of the Dehesa Comunal (Frontera, El Pinar). Livestock landscape unique in the Canary Islands, the Dehesa Comunal is a surviving remanant of a common grazing area established during the period of the Lordship in the old landowning regime. The area of the Dehesa Comunal covers some 4500 ha in the western part of the island. It corresponds to the steep slope on the leeward side that descends from 770 m above sea level to the sea.

Fig. 5 Tropical pineapple cultivation was the last significant crop to be incorporated into the agricultural landscape of El Hierro barely half a century ago

The distribution of El Cres and the Colonization Plan of the Dehesa Comunal that was approved in 1943 by the government of Franco's dictatorship brought some changes in this unique livestock landscape. The new distribution affected 300 ha of the Dehesa Comunal. This area corresponds to the higher, wetter and deeper soils of the Dehesa Comunal. El Cres was also divided into plots and given to the farming families of the region. After a few years of cultivation, the agricultural activity disappeared, but today the walls separating the plots and some livestock activities are still preserved (Martín 2006).

In the rest of the Dehesa Comunal, the Colonization Plan also ended up failing, as only a few infrastructures were carried out (small reservoirs, access tracks, parceling with stone walls, small reforestations). As in El Cres, the remains of the walls separating the plots built in the middle of Franco's colonization policy can still be seen today in the landscape, which is abandoned or underused (Martín 2006).

12. Livestock landscape of the Nisdafe plateau (Valverde). The deforestation of the green woodland (fayal-brezal and laurisilva) and its subsequent ploughing in the seventeenth century (Galván 1997) gave rise to this unique landscape of livestock farming. The basalt walls of the enclosures and the cattle trails, built to control the livestock, still mark today the vision of this landscape whose pastures are in an advanced state of abandonment.

13. Landscape of the tagasaste shrub-grassland plantations of the Isora "*lomadas*" (flat surfaces between ravines) (Valverde). Although they are distributed all over the island, these plantations of tagasaste (and tedera - Bituminaria bituminosa-, to a lesser extent) preserve a dominant landscape unit in these "*lomadas*". These plantations were probably started at the beginning of the twentieth century, but it was not until the middle of the twentieth century that they reached their maximum size in the middle of Franco's colonization policy. Today, they have been abandoned, although there are many livestock owners on the island who use this legume in times of scarcity of natural pastures or prolonged annual droughts.

14. Livestock-forest landscape of the green woodland of Fayal-Brezal ("*monte de dentro*") (Frontera, El Pinar). There is recent oral evidence of the use of the Monteverde for livestock in the areas where it is still preserved (Lorenzo 2011; Sánchez 2018). Sheep, but also goats and pigs took advantage of the grass growing under the forest canopy. Today, this practice has almost disappeared.

15. Livestock-forest landscape of the pine forests at the summit ("*monte de fuera*") (El Pinar). Similarly, the herbaceous species that grow under the island's pine forests have been used for livestock since time immemorial. Today, it is in disuse, but it is not uncommon to see sheep and goats in the forest of the leeward peaks.

Paleo-landscapes and agrarian landscape enclaves

This categorization includes paleo-landscapes or extinct agrarian landscapes and agrarian landscape enclaves or those with minimal current extension, since elements already in ruins can be observed or some residual extensions of them are still preserved. They are interesting because they provide information on the richness of the island's agro-cultural heritage. Table 1 contains a description of these extinct and/or reduced enclaves.

Local Informants: Acknowledgements Javier Morales (Agricultural Engineer-Head of Agriculture and Fisheries Service of the Cabildo Insular de El Hierro). Francisco Febles (organic livestock farmer of El Pinar). Alfredo Hernández (Technician of the Protected Designation of Origin of Wines "El Hierro"). Jennifer Quintero Melián (Cooperative Manager El Hierro Frontera).

References

Felipe JR, Herrera MP (2019) Plan Hidrológico de El Hierro. Consejo Insular de Aguas de El Hierro, El Hierro. http://www.aguaselhierro.org/planificacion/plan/plan2015

Fernández E, Monturiol F, Gutiérrez F (1974) Distribution and characterization of Canarian soils. II. Island of El Hierro. Ann Soil Sci Agrobiol XXXIII(5–6):358–370

Galván JA (1997) La identidad herreña. Centro de la Cultura Popular Canaria, La Laguna. ISBN: 84-7926-265-6

Hernández J, Niebla Tomé E (1985) El Hierro. In: Afonso L Geografía de Canarias. Geografía Comarcal, volume IV. Edirca, Santa Cruz de Tenerife, pp 145–180

Lorenzo MJ (2011) Communal lands and pastoral institutions on the island of El Hierro. Government of the Canary Islands, Canary Islands. ISBN: 978-84-606-5195-6

Martín CS (2006) Política territorial del franquismo en El Hierro (1940–1970). Idea Ediciones, Santa Cruz de Tenerife. ISBN: 84-96740-04-8

Sánchez S (2018) The move to the valley of El Golfo. Isla El Hierro. Le Canarien Ediciones, La Orotava (Tenerife). ISBN: 978-84-17522-14-8

Geomorphosites of El Hierro UNESCO Global Geopark (Canary Islands, Spain): Promotion of Georoutes for Volcanic Tourism

Javier Dóniz-Páez and Rafael Becerra-Ramírez

Abstract

Geotourism is a relatively recent concept and a novel kind of tourism, which has acquired a significant boom in the last decades, associated with the creation and consolidation of the UNESCO global geoparks network. There are two approaches to geotourism, one geological and the other geographical, much more global and inclusive of the elements of the natural and cultural heritage. In this chapter, we have chosen to use the geographical approach of geotourism, to apply it to the El Hierro global geopark and diversify the island's tourism, traditionally focused on diving and hiking, through geoforms (volcanic and non-volcanic), and its link with cultural heritage. For this, the most representative, preserved and accessible geomorphosites in the geopark have been identified, inventoried and selected, which can be visited through volcano tourism georoutes. To do this, a route is proposed in the El Faro-Orchilla geozone (GZH-07) of the geopark, since it is one of the best examples of recent monogenetic basaltic volcanism in the Canary Islands. This geozone has a high geodiversity and richness in its natural and cultural heritage, it is easily accessible and different geoforms and views of the island landscape can be seen along a 9.5 km route and 8 stops.

J. Dóniz-Páez (✉)
Geoturvol-Department of Geography and History, University of La Laguna, San Cristóbal de La Laguna, Spain
e-mail: jdoniz@ull.edu.es

R. Becerra-Ramírez
Geovol-Department of Geography and Territorial Planning, University of Castilla-La Mancha, Ciudad Real, Spain
e-mail: rafael.becerra@uclm.es

J. Dóniz-Páez · R. Becerra-Ramírez
Volcanological Institute of the Canary Islands (INVOLCAN), Granadilla de Abona, Spain

Keywords

El Hierro geopark • Geomorphosites • Geotourism • Volcano tourism

1 Introduction

Tourism, before the Covid-19 pandemic, has been one of the main global economic activities and its growth unstoppable (UNWTO 2019). Despite this, many mature sun and beach tourism destinations were already showing signs of decline before the current health crisis. This led to decreases in tourist arrivals and lack of renovation of tourist facilities and infrastructures (Hernández and Santana 2010), which, in turn, has caused stagnation and decline in these destinations (Simancas et al. 2020). Therefore, destinations require innovation and the creation of new tourism products and experiences. Consequently, in some destinations, new and sustainable tourism proposals are being developed such as geotourism or volcanic tourism (Dóniz-Páez et al. 2011, 2020a). Moreover, the diversity of geoforms in volcanic areas are very attractive (Németh et al. 2017) for visitors and especially for those interested in geotourism (Erfurt-Cooper 2018).

Geotourism has gained some momentum in recent decades (Dowling 2013; Dowling and Newsone 2018) as it has been associated with the creation and consolidation of a global network of geoparks. Geotourism is a relatively recent concept and a novel tourism modality (Pásková and Zelenka 2018) in which two types of approaches can be recognized: geological or geographical (Dowling and Newsone 2018), with the geographical approach being much more global and integrative (Tourtellot 2000; National Geographic 2010; Dóniz-Páez et al. 2020a, b). This has led to the development of geotourism initiatives, products and experiences, which though different from each other do not have to be mutually exclusive (Dowling and Newsone 2018).

In the geopark of El Hierro, the geoforms directly condition the rest of the natural and cultural heritage. For this reason, in this study, we have opted for a geographical approach to geotourism, as it is much more in line with the reality of El Hierro. In this sense, the aim of this study is to identify, classify and select representative, preserved and accessible geomorphosites of El Hierro UNESCO Global Geopark. This will promote El Hierro's natural and cultural heritage and contribute to diversifying tourism on the island, currently focused on diving and hiking (Dóniz-Páez et al. 2011). It will also boost the economic development of El Hierro through the creation of volcanic tourism georoutes as seen in other volcanic areas aspiring to become geoparks (Becerra-Ramírez et al. 2020).

2 Methodology

Following Bouzekraoui et al. (2017), the methodology used in this work consists of three stages: identification, classify and selection of volcanic geomorphosites of El Hierro's geopark for tourism georoutes. The identification of geomorphosites was performed by topographic, geological and geomorphological mapping at different scales, Dems of El Hierro and field work by members of the research team over the last 20 years on the island (Dóniz-Páez et al. 2011). For the classification of the geomorphosites, we used the methodological proposal by Dóniz-Páez et al. (2020b) on the diversity of volcanic geoheritage in the Canary Islands. In the selection of the volcanic geomorphosites, it was considered that they should be representative of the geomorphological diversity of the geopark of El Hierro. Moreover, they should be well preserved and accessible via the current road network of the island or through the maritime routes that diving companies usually take.

3 Results and Discussion

The island of El Hierro is the smallest, geologically the youngest and geographically the westernmost of the Canary Islands. El Hierro is of volcanic genesis, and the oldest materials are from a million years ago. The island was formed from the polygenetic edifices of Tiñor (1.12–0.88 Ma), El Golfo-Las Playas (545–176 ka) and Rift volcanism (158 ka-present) (Becerril et al. 2016; Aulinas et al. 2019). The general morphology is in the shape of a three-pointed star resulting from the seafloor fractures on which it has been built following three rifts (Carracedo 2008). In general, the volcanic materials of El Hierro are a succession of basaltic lavas, agglomerates of volcanic tuffs

and monogenetic volcanoes (Carracedo 2008). It is still a volcanically active island complex, whose last eruption was underwater and occurred between 2011 and 2012 in the Sea of Calms, in the south of El Hierro. The island has also been subjected to the processes of erosion and accumulation giving rise to outstanding geomorphological landscapes.

3.1 Geomorphosites of El Hierro Island

The classification of the geomorphosites of El Hierro corresponds to the proposal of Dóniz-Páez et al. (2020b). This groups the geomorphosites into volcanic landforms and processes or non-volcanic landforms and processes. Within the first group, they are divided into magmatic and hydromagmatic volcanic cones (monogenetic, polymagmatic and polygenetic), terrestrial and submarine lava flows (lava delta, pahoehoe, aa, blocks) and other volcano geoforms (dykes). Regarding non-volcanic landforms, the classification includes recent landforms, giant landslides and relict or fossil landforms.

The island has 230 volcanic cones amounting to 0.8 cones/km^2 (Becerril et al. 2016). They are monogenetic basaltic volcanoes of magmatic (Hawaiian, Strombolian and violent Strombolian and Vulcanian dynamics) and hydromagmatic dynamics, built by lapilli, scoria, spatter and lavas of diverse morphology with open, closed and multiple craters. Although most of the volcanic cones are monogenic, some examples of polymagmatic volcanism, such as Tanganasoga can be recognized (Dóniz-Páez et al. 2020b). These volcanoes produced large terrestrial lava flows of varied morphology of pahoehoe, aa and blocks with the formation of lava deltas (Tamaduste, Tacorón…) and some of the most diverse lava fields of the Canary Islands (Los Lajiales) (Beltrán Yanes and Dóniz-Páez 2009). However, in some areas of the island, examples of submarine lavas can be observed, such as in La Caleta. Although the greatest diversity of volcanic geoheritage corresponds to volcanic cones and lava flows, in El Hierro Geopark, other volcanic (dikes) and non-volcanic (San Andres fault) geoforms can be found.

El Hierro is a geologically young island in which the volcanic landforms are very important, but non-volcanic landforms and processes are also present. Among these recent landforms, we can mention the cliffs, the ravines, alluvial and colluvial deposits and several red (Hoya Verodal) and black sand beaches. In relation to relict or fossil non-volcanic landforms, they are not very common though some beaches and small dunes in Arenas Blancas stand out. However, the giant landslides such as those of Tiñor, El Golfo, Las Playas and El Julan (Carracedo 2008) are the geoforms that best define the current morphology of El Hierro.

3.2 Selection of Geomorphosites

Taking into account the diversity of the geomorphological heritage identified in the previous section, the most representative, best preserved and most accessible geomorphosites of El Hierro can be selected. This is important when designing itineraries and georoutes for volcanic tourism. In this sense, according to the classification of landforms and relief processes above. Table 1 shows the main geomorphosites of each of the categories defined according to Dóniz-Páez et al. 2020b.

3.3 Georoutes for Volcanic Tourism

The diversity of direct volcanic and non-volcanic landforms and processes in El Hierro's geopark means multiple georoutes can be created for volcanic tourism. Depending on the criteria chosen, specific georoutes can be created according to certain features, such as basaltic cinder cones, hydromagmatic volcanoes, recent lava fields, giant landslides, coastal landforms, etc. However, volcanic tourism georoutes can also be chosen according to a broader perspective that encompasses a wide variety of elements of the natural and cultural heritage of the geopark (Fig. 1).

In this work, we have decided to develop a volcanic tourism georoute according to the geographical approach of geotourism. The place chosen for the itinerary is the El Faro-Orchilla-lava flow geozone (GZH-07) of the geopark. The selection of Orchilla lava delta is due to the richness of its natural and cultural heritage and its easy accessibility

(Dóniz-Páez et al. 2019). The diversity of this geozone is associated with the presence of one of the best examples of recent monogenic, basaltic magmatic volcanism in the geopark, where different types of volcanic edifices can be identified (cinder cones, spatter cones, hornitos, etc.) built by lapilli, spatter, lavas, bombs, etc. and with varied shapes (Ring-shaped cones, horseshoe-shaped cones, multiple scoria cones, volcano without crater, etc.), spectacular lava fields with pahoehoe, aa and block morphologies with lava tubes and channels and accretion balls. In addition to the volcanic forms, other morphologies such as fossil and active cliffs, beaches, ravines and important alluvial fans can be observed. In this geozone, the cultural heritage is associated with the volcanic heritage through the use of the volcanic tubes and *jameos* (volcanic caves) by the population for residences or livestock huts, quarries for the extraction of lapilli, various infrastructures associated with the use of water and, above all, the presence of the Orchilla Lighthouse and the Monument to the Zero Meridian, which are two tourism icons of El Hierro. It is also worth mentioning the impressive panoramic views of the whole area of El Julan, one of the four giant landslides in El Hierro. Furthermore, this is the starting point of the GR-131 trail that runs through the whole of the Canary Islands. For all these reasons, the geopark stands out for its interesting local values and for its scientific, didactic and tourism interest (Table 2; Fig. 2).

The volcanic georoute has eight stops and a total distance of 9.5 km divided into two parts (Fig. 3). The main itinerary runs along the HI-503 La Montaña road towards the Virgen de Los Reyes hermitage in La Dehesa and continues along a detour towards the Orchilla Lighthouse for 6.5 km. The

Table 1 Main geomorphosites selected of El Hierro geopark	Volcanic landforms	Main geomorphosites selected
	Volcanic cones	Monogenetic: magmatic (Corona del Lajial, Lomo Negro, Orchilla Geozone, Escobar, Chamuscada, etc.), hydromagmatic (Hoya Fileba, Ventejis and Hoya Verodal) and submarine eruption of Tagoro volcano Polymagmatic: Tanganasoga Polygenetic: Tiñor and El Golfo ediffices and volcanic rifts
	Lava flows	Terrestial: pahoehoe (Los Lajiales, Orchilla, Bahía de la Hoya, Tamasina, Calcosas), aa (Orchilla, Timijiraque, Tamasina and Tacorón), blocks (Tamaduste and Orchilla) Submarine: La Caleta pillow lavas
	Others	Dykes
	Non-volcanic landforms	
	Recent	Ravines: Gorreta, Barranco de los Trabaditos, Barranco del Jable, Bascos Valley, Barranco de Tejeda, Barrancos de Las Playas giant landslides, Barranco de las Playecitas, Barranco del Balón, Barranco del Tiñor Beaches: black (Timijiraque and Las Playas) and red (Hoya Verodal and Playa de la Arena in Tacorón) volcanic sand Taluses: Gulf and Beach deposits
	Giant landslides	El Golfo, Las Playas and El Julan
	Fossil	Arenas Blancas beach, La Caleta
	Others	San Andrés fault

Source Prepared by authors

Fig. 1 Several geomorphosites selected in El Hierro UNESCO Global Geopark: Hornito in Lomo Negro cinder cone (**a**); Chamuscada monogenetic basaltic volcano (**b**); Hoya Verodal tuff ring (**c**); Hoya Fileba hydromagmatic volcano (**d**); Tanganasoga polymagmatic volcano (**e**); Tiñor ediffice (**f**); Pahoehoe lava flows in Bahía de La Hoya (**g**); Aa lava flow in Tacorón (**h**); La Arena volcanic red sand beach (**i**); Talusses in El Golfo (**j**); Las playas giant landslides (**k**); Fossil sand in La Caleta (**l**)

Table 2 Natural and cultural volcanic heritage of Orchilla geozone georoute

Stops	Natural heritage	Cultural heritage
1. Hoya Bajo – Caldereta del Tabaibal Manso	Landforms of the geozone, recent volcanism and coastal shrubs	Camino de la Virgen, Orchilla Lighthouse, quarries, trails
2. Montaña Toscones volcano	Volcanoes, craters, lava flows, ravines, taluses, *Rumex lunaria* and *Euphorbia lamarckii* shrubs	Cistern
3. Hornito – Montaña Calcosas volcano	Cinder-Scoria cones, nested cinder-scoria cones hornitos, spatter cones, craters, pahoehoe and aa lavas, lava lakes, lava tubes, tube subsidence (*jameos*), lava channels, gullies, *Euphorbia sp.* shrubs, birds	Cabins in tube subsidence (jameos)
4. Eruptive fissure	Hornitos and pahoehoe lavas and tubes, *jameos* and channels, lapilli, gullies, coastal shrubs	Remains of a telephone installation in *jameos*
5. Paleocliff – Orchilla volcano	Slope break, cinder-scoria cone, spatter cones, hornitos, colluvions, alluvial fans, rupicolous vegetation on recent lavas	Quarries, trails, wooden signs
6. Monument to the Meridian 0	Alluvial fans, cinder-scoria cones, aa lavas, cliffs, *Euphorbia sp.* and salty shrubs	Zero Merididan Monument
7. Orchilla Lighthouse	Pahoehoe lavas, lava tubes and channels, *jameos*, cinder-scoria cones, spatter cones, cliff, El Julan landscape view, Las Calmas sea, whales watching	Orchilla lighthouse, cabins, dry-stone walls, cattle fold (*gorona*), trails, Cross
8. Montaña Negra volcano - La Laja de Orchilla	Cinder-scoria cones, aa lavas, lava tubes and channels, rocky shore, abrasion platform, El Julan and cliff views	Dock and leisure area

Source Prepared by authors

Fig. 2 Key stops on the volcanic georoute in the Orchilla Geozone GZH07. The numbers refer to Table 2

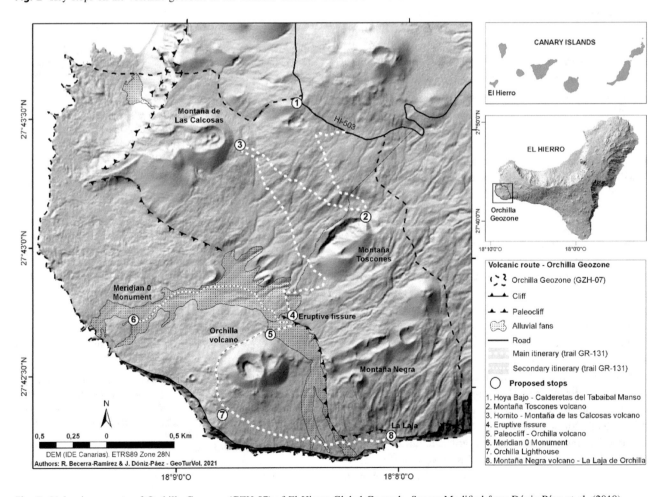

Fig. 3 Volcanic georoute of Orchilla Geozone (GZH-07) of El Hierro Global Geopark. *Source* Modified from Dóniz-Páez et al. (2019)

secondary itinerary turns off at stop 5 at the Orchilla volcano along the HI-504 track and heads towards the Zero Meridian Monument. It is a round trip of about 3 km. The maximum elevation is 464 m at stop 1 and the minimum 0 m at stop 8, which is the pier of La Laja de Orchilla. Although the entire georoute can be done by car, it is recommended that stop 5–stop 6 of the Monument to the Zero Meridian and return is completed on foot or bicycle because of the difficulties there may be for road traffic, since it runs through the deposits of alluvial fans.

4 Conclusions

Although it is true that the current health crisis has contributed to diversifying the tourism offer in many regions, before this pandemic, many destinations had already created new tourism products and experiences based on the promotion of local heritage. In the case of geotourism, for some years now, initiatives have been developed around two approaches (geological and geographical) that are complementary to each other. Indeed, volcanic tourism has been a reality in volcanic areas for some years now. Even so, it can be said that geotourism is still a relatively recent, innovative and sustainable form of tourism, closely linked to the management of geoparks. In this sense, the island of El Hierro has opted for a tourism distinct from the Canary Islands as a whole, one based on sustainability and the promotion of local geographical heritage and mainly focused on hiking and diving. Therefore, this paper has evaluated the volcanic geoheritage on the island, which led to the creation of a geopark in 2014 following the underwater eruption of the Tagoro volcano in 2012. For this purpose, different volcanic and non-volcanic geomorphosites have been identified, classified and selected to be representative, preserved and accessible examples of the geodiversity of El Hierro and of its cultural heritage. These geomorphosites are incorporated in geotourism georoutes from a geographical approach in the geopark, increasing and diversifying the tourism offer on the island. The Orchilla geozone has been selected for the proposed volcanic tourism georoute based on the diversity of its volcanic forms (volcanic cones and lava flows), those from erosion and accumulation (landslides, ravines, cliffs, deposits, beaches, etc.) and the interesting cultural heritage linked to a historically inhospitable region with a semi-arid climate. We must also mention the good accessibility, the existence of several approved trails, the presence of a bathing and leisure area (Las Lajas) and two of the most significant tourist icons of El Hierro (the Orchilla Lighthouse and the Zero Meridian Monument). In this work, we have developed a model georoute for volcanic tourism in the geopark, however, depending on the geomorphosites identified and selected, many other georoutes can be developed.

Acknowledgements This research was supported by project "VOLTURMAC, Fortalecimiento del volcano turismo en la Macaronesia (MAC2/4.6c/298)", and is co-financed by the Cooperation Program INTERREG V-A Spain-Portugal MAC (Madeira-Azores-Canarias) 2014–2020, http://volturmac.com/.

References

Aulinas M, Domínguez D, Rodríguez-González A, Carmona H, Fernández-Turiel J, Pérez Torrado F, Carracedo J, Arienzo I, D'Antonio M (2019) The Holocene volcanism at El Hierro: insights from petrology and geochemistry. Geogaceta 65:35–38

Becerra-Ramírez R, Gosálvez RU, Escobar E, González E, Serrano-Patón M, Guevara D (2020) Characterization and geotourist resources of the Campo de Calatrava Volcanic Region (Ciudad Real, Castilla-La Mancha, Spain) to develop a UNESCO global geopark project. Geosciences 10(11):441. https://doi.org/10.3390/geosciences10110441

Becerril L, Galve J, Morales J, Romero C, Sánchez N, Martí J, Galindo I (2016) Volcano-structure of El Hierro (Canary Islands). J Maps 12(1):43–52. https://doi.org/10.1080/17445647.2016.1157767

Beltrán Yanes E, Dóniz-Páez J (2009) 8320, campos de lava y excavaciones naturales. In: Ministerio de Medio Ambiente, y Medio Rural y Marino (eds) Bases ecológicas preliminares para la conservación de los tipos de hábitat de interés comunitario en España. Ministerio de Medio Ambiente, y Medio Rural y Marino, Madrid, pp 1–124. Retrieved from https://www.researchgate.net/publication/291345482_8320_Campos_de_lava_y_excavaciones_naturales

Bouzekraoui H, Barakat A, Touhami F, Mouaddine A, El Youssi M (2017) Inventory and assessment of geomorphosites for geotourism development: a case study of Aït Bou Oulli valley (Central High-Atlas, Morocco). Area 50:331–343. https://doi.org/10.1111/area.12380

Carracedo J (2008) Los volcanes de las Islas Canarias. IV. La Palma, La Gomera, El Hierro. Rueda, Madrid, ISBN 978-8472071902

Dóniz-Páez J, Becerra-Ramírez R, González E, Guillén C, Escobar E (2011) Geomorphosites and geotourism in volcanic landscapes: the example of La Corona del Lajial Cinder cone (El Hierro, Canary Islands, Spain). GeoJ Tourism Geosit IV(2):185–197. Retrieved from http://gtg.webhost.uoradea.ro/PDF/GTG-2-2011/3_98_Doniz_Paez.pdf

Dóniz-Páez J, Becerra-Ramírez R, Anceaume-Chinea L (2019) Ruta volcánica en el geoparque mundial Unesco de El Hierro (Canarias, España): geozona de Orchilla. In: Martín González E, Coello Bravo J, Vegas J (eds) Actas de la XIII Reunión Nacional de la Comisión de Patrimonio Geológico. IGME, Madrid, pp 111–116. Retrieved from https://www.researchgate.net/publication/333966261_Ruta_volcanica_en_el_geoparque_mundial_Unesco_de_El_Hierro_Canarias_Espana_geozona_de_Orchilla

Dóniz-Páez J, Hernández P, Pérez NM, Hernández W, Márquez A (2020a) TFgeotourism: a project to quantify, highlight, and promote the volcanic geoheritage and geotourism in Tenerife (Canary Islands, Spain). In: Németh K (ed) Volcanoes—updates in volcanology. Intechopen. https://doi.org/10.5772/intechopen.93723

Dóniz-Páez J, Beltrán Yanes E, Becerra-Ramírez R, Pérez NM, Hernández P, Hernández W (2020b) Diversity of volcanic geoheritage in the Canary Islands, Spain. Geosciences 10:390. https://doi.org/10.3390/geosciences10100390

Dowling R (2013) Global geotourism: an emerging form of sustainable tourism. Czech J Tour 2(2):59–79. Retrieved from http://www.czechjournaloftourism.cz/cislo/en/102/02-2013/#clanek104

Dowling R, Newsome D (2018) Geotourism: definition, characteristics and international perspectives. In: Dowling R, Newsome D (eds) Handbook of geotourism. Edward Elgar, Cheltenham, pp 1–22

Erfurt-Cooper P (2018) Geotourism development and management in volcanic regions. In: Dowling R, Newsome D (eds) Handbook of geotourism. Edward Elgar, Cheltenham, pp 152–167

Hernández R, Santana A (2010) Destinos turísticos maduros ante el cambio. Reflexiones desde Canarias. I.U. Ciencias Políticas y Sociales, Universidad de La Laguna, La Laguna. ISBN 978-8461433865

National Geographic (2010) What is geotourism? Center for Sustainable Destinations. Retrieved from https://www.nationalgeographic.com/maps/article/about-geotourism

Németh K, Casadevall T, Moufti MR, Marí J (2017) Volcanic geoheritage. Geoheritage 9:251–254. https://doi.org/10.1007/s12371-017-0257-9

Pásková M, Zelenka J (2018) Sustainability management of UNESCO global geoparks. Sustain Geosci Geotour 2:44–64. Retrieved from https://www.scipress.com/SGG.2.44

Simancas Cruz M, Hernández Martín R, Padrón Fumero N (eds) (2020) Turismo pos-COVID-19. Reflexiones, retos y oportunidades. Cátedra de Turismo CajaCanarias-Ashotel de la Universidad de La Laguna, La Laguna. ISBN 978-8409218165

Tourtellot JB (2000) Geotourism for your community. National Geographic Drafts, Washington DC, p 2

UNWTO (2019) UNWTO data dashboard. https://www.unwto.org/es/unwto-tourism-dashboard

Birdwatching as a New Tourist Activity in El Hierro Geopark

Rafael Ubaldo Gosálvez Rey, Adrián Navas Berbel, and Diego López de la Nieta González de la Aleja

Abstract

Bird watching is one of the most popular ways of getting close to nature, laying the foundations for what is now known as Birdwatching or Birding, nowadays a niche within ecotourism. The Canary Islands are an exceptional centre for ornithological tourism, standing out for the presence of six endemic species that are exclusive worldwide. In this context, the island of El Hierro is the least visited island for bird watching in the Canary Islands archipelago, even though it has been designated as a Biosphere Reserve and Geopark. This paper aims to lay the foundations for the development of ornithological tourism on the island of El Hierro, following the methodology proposed by Gosálvez Rey (El Valle de Alcudia y Sierra Madrona, 2009), Puhakka et al. (PLoS One 6, 2011) and the Ornithological Tourism Strategy for the Canary Islands (SEO/Birdlife in Estrategia de Turismo Ornitológico para la Macaronesia, 2016). A geographical analysis of the diversity of species is addressed, the most suitable trails and points for birdwatching are indicated and the mechanisms for promoting this tourist activity are outlined. The island of El Hierro has 22 species and subspecies of birds of interest for birdwatching, the best areas for birdwatching being the Natura 2000 sites (EU) and Birdlife International's IBAs. The island of El Hierro is served by a network of paths provided by the Cabildo de El Hierro and the Spanish government's Caminos Naturales programme, complemented by a set of fourteen lookout that serve as strategic points for bird watching. The challenge for the island of El Hierro will be to develop birdwatching that respects and even enhances natural values, avoiding endangering the species observed and their habitats.

Keywords

Birdwatching • Conservation • El Hierro • Ecotourism • Birds

1 Introduction

Of the various options for humans to approach nature, birdwatching is one of the most popular. Birdwatching started in the late nineteenth century as an alternative to scientific hunting and developed thanks to public awareness campaigns and the emergence of ornithological organizations (Szczepańska et al. 2014). United Kingdom and United States were pioneers in birdwatching and in the creation of societies, such as the Royal Society for the Protection Birds in 1889 and the National Audubon Society in 1905, to the point that birdwatching became a main leisure activity for large segments of their populations (López Roig 2008). Private entrepreneurs saw an opportunity in its commercialization laying the foundations of what is now called "ornithological tourism", "avitourism", birdwatching or birding (Jones and Buckley 2001), today considered as a niche within nature tourism (Şekercioğlu 2002; Biggs et al. 2011). Ornithological tourism has been defined by Rivera (2007) as an activity that involves travelling from a place of origin to a specific destination with the aim of observing the local avifauna in their natural environment, providing economic benefits for destinations.

R. U. Gosálvez Rey (✉)
GEOVOL-IHE, Department of Geography and Territorial Planning, University of Castilla-La Mancha (UCLM), Ciudad Real, Spain
e-mail: RafaelU.Gosalvez@uclm.es

A. Navas Berbel · D. López de la Nieta González de la Aleja
Department of Geography and Territorial Planning, University of Castilla-La Mancha (UCLM), Ciudad Real, Spain
e-mail: Adrian.Navas@uclm.es

D. López de la Nieta González de la Aleja
e-mail: diegolopezdelanieta@gmail.com

R. U. Gosálvez Rey
Volcanological Institute of the Canary Islands (INVOLCAN), Granadilla de Abona, Spain

© The Author(s) 2023
J. Dóniz-Páez and N. M. Pérez (eds.), *El Hierro Island Global Geopark*, Geoheritage, Geoparks and Geotourism, https://doi.org/10.1007/978-3-031-07289-5_9

In the last twenty years, ornithological tourism has experienced a boost in Spain from private and public initiatives, especially projects linked to European funds for rural development (Gosálvez Rey 2009). These initiatives aim to generate a specialized economic sector that includes guide services, accommodation, restaurants, car rental or construction of infrastructures such as observatories and visitors' centres.

Currently, our country is one of the main birdwatching destinations for ornithological tourists, representing 10% of ornithological tourism packages that British operators have in their sales catalogues worldwide (SEO/Birdlife 2016), which demonstrates the potential that Spain has for ornithological tourism. The most consolidated destinations for this activity in Spain are Andalusia, especially Doñana, Campo de Gibraltar, Extremadura and the Pyrenees.

In this context, the Canary Islands could become an exceptional centre for ornithological tourism given their mild climate, splendid landscapes and good transport links. The ornithological interest exists as they are home to some 90 breeding species and more than 380 that visit the islands throughout the year (AVIBASE), with six exclusive endemic species worldwide: Bolle's Pigeon, Laurel Pigeon, Fuerteventura Stonechat, Tenerife Blue Chaffinch, Canary Islands Chiffchaff and African Blue Tit. These give the Canary Islands the richest avifauna of all the Macaronesian archipelagos (SEO/Birdlife 2016). This has led BirdLife International to consider the Canary Islands as an EBA (Endemic Bird Area), thus making it the only EBA in Western Europe. Despite this ornithological strength, this tourism product has very little weight in the tourism market of the archipelago, where sun and beach tourism are still the leaders (SEO/Birdlife 2016). In fact, there are only two privates companies dedicated to birdwatching, Birding Canarias and Aves Ecotours, which are struggling to make progress.

The island of El Hierro is the furthest away from the African continent and the smallest of the Canary archipelago, despite its small size, it contains the whole range of habitats present in the Canary Islands, except for the scrubland of summits. There are areas of 'Monteverde' (forested hills), Canary Island pine and juniper woodlands where it is relatively easy to observe endemic bird species and subspecies of the archipelago. Birds of open farmland, birds of prey, corvids and important numbers of gulls and migrant birds can be observed in La Dehesa and on the Llanos de Nisdafe. Finally, seabirds can be observed from the Orchilla lighthouse, La Restinga, the Roques de Salmor or in the Bahía de Naos-Hoya de Tacorón. Despite this, it is the least visited island for birdwatching in the Canary archipelago, even though the whole island has been designated by UNESCO as a Biosphere Reserve and Geopark.

This work aims to lay the foundations for the development of ornithological tourism on the island of El Hierro: an area of great interest for birdwatching where, to date, 178 species of birds have been indexed (AVIBASE), of which almost fifty breed regularly on the island. To do this, a geographical analysis of the diversity of species is conducted indicating the main taxa of tourist interest; the places and routes that provide the best spots for birdwatching and the mechanisms available to promote this tourist activity without affecting the conservation of the birds and their habitats.

2   Methods

We followed the methodology proposed by Gosálvez Rey (2009), Puhakka et al. (2011) and the action programme contained in the Ornithological Tourism Strategy for Macaronesia in the Canary Islands promoted by SEO/Birdlife (2016).

The geographical analysis of the distribution of bird species on El Hierro has considered three aspects: total species richness, number of endemic taxa and number of endangered species. This makes it possible to determine the "target" species that would attract an ornithological tourist to visit El Hierro. For this purpose, the information contained in the atlases of breeding and wintering birds of Spain (Martí and Del Moral 2003; SEO/Birdlife 2012) and in the Canary Islands Natural Inventory Bank (BIOCAN) has been considered. For the location of places of interest for birdwatching on El Hierro, the SPAs-Natura 2000 network and the Important Birds Area (IBA) network (Viada 1998) were used as starting points. Information contained in the websites of birdwatching companies and blogs (Reservoir Birds, BirdForum, juanjoramoseco.com) was also consulted. The network of trails and paths of El Hierro (Hikes of El Hierro and Traditional Paths of El Hierro Nature Trail) serves as a basis for designing ornithological tourist packages based on the distribution of birds and those areas where it is more likely to observe them, thus identifying the areas with the greatest potential for birdwatching. The free software QGIS (version 3.16.1) has been used to consult all the spatial data and generate maps, data related to the objectives of this study are shown in Fig. 1.

3   Is It Worth Travelling to El Hierro for Birdwatching? Geographical Analysis of Bird Diversity

Birds, like other fauna, are intimately conditioned by the factors of the ecological environment in which they live. Relief, climate and water, vegetation and human activity are the main factors that explain the distribution of the different species of birds on the island of El Hierro.

Fig. 1 Data used in the study and their relationship to the research objectives. *Source* Gosálvez Rey (2009), Puhakka et al. (2011)

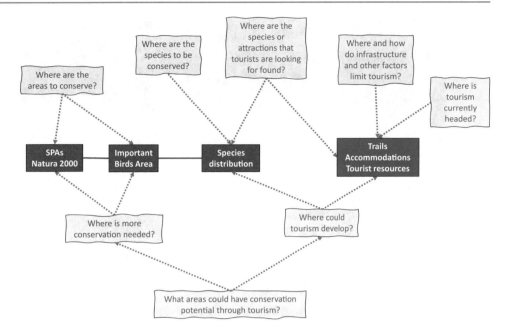

Carrascal and Palomino (2002) analysed inter-island variations in the number of nesting land bird species in the Canary and Savage archipelagos, concluding that variations in species richness increased with increasing island size and greater habitat diversity and decreased with increasing distance from the mainland. The island of El Hierro is therefore the island with the lowest value of nesting bird richness in the Canary archipelago.

However, one of the simplest measures used to estimate the biological diversity of a territory is the richness or number of species which, although it is not a good indicator of the complexity of diversity, is easy to obtain and is often used as a first approximation (Margalef 2005; Lomolino et al. 2016). The species richness of birds presents on the island of El Hierro as of 2021 is 178 taxa according to AVIBASE. Of these, 31 species belong to the phenological category of rare or accidental, 42 are breeding species and the rest are migrants or habitual winterers (Martí and Del Moral 2003; SEO/Birdlife 2012). Of these 178 species, 21 are included in the Canary Islands catalogue of protected species, with the Canary Island Raven as the only species in danger of extinction and the Osprey, Manx Shearwater, Bolle's Pigeon and Laurel Pigeon as vulnerable. The rest are included in the category of species of interest for the Canary Island ecosystems or in the category of special protection.

The spatial distribution of birds on El Hierro responds to the variety of habitats present on the island, a variety that is dependent on two main factors: the altitudinal gradient, with geomorphological, climatic and biogeographical implications, and the action of human activity that has developed on this island. There are currently six main types of habitats for birds (Fig. 2): forest areas (Canary Island pine forests,

Monteverde and juniper woodlands); open spaces and extensive agrosystems (coastal scrubland, grasslands and crops); cliffs, islets, rocks and volcanic Badlands; the coastal strip and its beaches; artificial wetlands (reservoirs and artificial ponds); and urban areas. A detailed description of them is beyond the scope of this paper, so we recommend consulting their characterization in Martín and Lorenzo (2001), in Fernández-Palacios et al. (2001) and in del Arco Aguilar and Rodríguez Delgado (2018).

In this analysis, what we are interested in highlighting is whether there are birds of interest on this island to justify the arrival of ornithological tourists. Considering the area of distribution at European level, the category of threat and the phenological status of the birds present on El Hierro, a total of 21 species and subspecies of birds have been selected (Table 1; Fig. 2) for which an ornithological visit to this island is essential.

4 Areas and Trails for Bird Watching in El Hierro

For a territory to become a tourist destination for birdwatchers, not only must there be species of interest, there must also be places where it is relatively easy to observe birds, and there must be a level of infrastructure to cater for this tourist activity. At present, the island of El Hierro has these three elements: birds of interest, the possibility of visiting places where it is easy to watch birds (Fig. 3) and an infrastructure of roads, trails and lookouts.

The best areas for bird watching are those where there is a good chance of seeing the most emblematic species, while at

Fig. 2 Some species of interest for ornithological tourism on the island of El Hierro: **a** African Blue Tit, **b** Berthelot's Pipit, **c** Yellow-legged Gull, **d** Atlantic Canary, **e** Common Kestrel, **f** Canary Islands Raven. *Author photographs* Rafael Ubaldo Gosálvez Rey

Table 1 Bird species of interest for ornithological tourism on the island of El Hierro distributed by habitat

	Widely distributed habitats	Habitats of restricted distribution
Species of interest	**Forest areas**	**Cliffs, islets and rocky outcrops**
	Bolle's Pigeon *Columba bollii*	Red-billed Tropicbird *Phaethon aethereus*
	Laurel Pigeon *Columba junoniae*	White-faced Storm Petrel *Pelagodroma marina*
	Eurasian Sparrowhawk *Accipiter nisus granti*	Cory's Shearwater *Calonectris borealis*
	Common Buzzard *Buteo buteo insularum*	Bulwer's Petrel *Bulweria bulweria bulwerii*
	Common Chaffinch *Fringillia coelebs ombriosa**.	Osprey *Pandion haliaetus*
	African Blue Tit *Cyanis. teneriffae ombriosus**.	Barbary Falcon *Falco pelegrinoides*
	Canary Is. Chiffchaff *Phylloscopus canariensis*	Canary Islands Raven *Corvus corax canariensis*
	Eurasian Blackcap *S. atricapilla heineken*	Plain Swift *Apus unicolor*
	Goldcrest *Regulus regulus ellenthalerae*	
	Open spaces and agrosystems	**Coastal strip and beaches and Wetlands**
	Eurasian Stone-curlew *B. oedicnemus distinctus*	No species of interest
	Common Kestrel *Falco tinnunculus tinnunculus canariensis*	
	Berthelot's Pipit *Anthus berthelotii*	
	Atlantic Canary *Serinus canaria*	
	Urban spaces	
	Plain Swift *Apus unicolor*	
	Canary Islands Chiffchaff *Phylloscopus canariensis*	
	African Blue Tit *Cyanistes teneriffae ombriosus**.	
	Common Kestrel *Falco tinnunculus tinnunculus canariensis*	

Source: Prepared by authors

*Subpecies exclusive to the island of El Hierro

the same time not disturbing them excessively and, if possible, in a safe and orderly way. In the case of the island of El Hierro, there are two networks of areas of interest for birdwatching: the Special Protection Areas for Birds (SPAs-Natura 2000) of the European Union and the Important Bird Areas (IBAs) developed by Birdlife International. In addition to these areas, there are other European (SAC-Natura 2000) and regional protection organisations as well as ones of international recognition by UNESCO

(Biosphere Reserve and Geopark). Overall, 100% of the surface area of the island of El Hierro is covered by them (Table 2).

If we focus on SPAs, we should point out that they were created in 1979 following the approval of the Birds Directive, the first regulation issued by the European Union for nature conservation. The Birds Directive identifies 200 endangered taxa for which it is necessary to designate special protection areas. Since 1992, SPAs have been integrated

Fig. 3 Main habitats for fauna on the island of El Hierro: **a** Canary Island pine forest in El Julan, **b** Monteverde on the escarpment of El Golfo, **c** Juniper woodlands in La Dehesa, **d** Agrosystems in the Llanos de San Andrés, **e** Cliffs and Roques de Salmor, **f** Urban habitat, Valverde. *Author photographs* Rafael Ubaldo Gosálvez Rey

into Natura 2000, together with the Special Areas of Conservation (SACs), following the approval of the Habitats Directive. Natura 2000 constitutes of a network of protected areas to guarantee the conservation of biodiversity within the framework of the European Union. Three areas on El Hierro have been declared SPAs: Garoé, El Hierro and Gorreta and Salmor.

Garoé (ES10000102)

This area is in the north-eastern sector of El Hierro. It consists of the Macizo de Ventejís on whose windward slopes are the remains of the laurel forest that populated the whole area in the past. There is a peculiar agricultural landscape in the Llanos de San Andrés or Nisdafe, in which pastures and crops predominate. This area has great cultural value due to the magical character that the mountains had for the *bimbaches* (indigenous inhabitants), highlighting the famous Garoé, sacred tree for the *bimbaches* and symbol in the coat of arms of the island of El Hierro. The birds that justify the declaration of this area as a SPA are the Eurasian Sparrowhawk, Common Buzzard, Canary Common Chaffinch, Common Kestrel and Long-eared Owl.

El Hierro (ES0000103)

This SPA has an altitudinal range that goes from the coastline to 1501 m above sea level, which favours the appearance of different microclimates responsible, in turn,

for the main habitats that appear on the island. This wide range of habitats facilitates a great diversity of birds, identifying some 35 species, including seabirds, raptors and endemic passerines. Among the seabirds, Bulwer's Petrel, Cory's Shearwater, Barolo Shearwater, European Storm Petrel, Band-rumped Storm Petrel and Common Tern stand out. Raptors are represented by Osprey, Eurasian Sparrowhawk, Common Kestrel, Common Buzzard and Long-eared Owl. The passerines include the Canary Common Chaffinch (subsp. *ombriosa*), African Blue Tit (subsp. *ombriosus*), Atlantic Canary, Berthelot's Pipit, Canary Island Raven, Bolle's Pigeon and Laurel Pigeon.

Gorreta and Salmor (ES10000104)

Located in the north–north eastern part of the island of El Hierro, it is a large cliff with drops of up to 1000 m, in the easternmost part are the Roques de Salmor. The main importance of this area are the colonies of seabirds, where several species nest and reproduce such as Bulwer's Petrel, Cory's Shearwater, Barolo Shearwater, European Storm Petrel (with one of the largest national populations), Band-rumped Storm Petrel and Osprey. Other bird species of interest include Common Kestrel, Common Buzzard, Canary Island Raven, Western Barn Owl and Canary Island Chiffchaff.

The Important Bird Areas programme is an initiative of the European Commission, which arose to help compare national contributions to the lists of SPAs in the Birds

Table 2 Figures of protection of El Hierro

Figure	Code and name	Area (ha)	Habitats
SPA	ES0000102 Garoé	1124	Monteverde and agrosystems
	ES0000103 El Hierro	12,406	Cliffs, islets and rocks
	ES0000104 Gorreta and Salmor	595	
Total		14,125	
IBA	385 Macizo de Ventejís	1289	Monteverde and agrosystems
	386 Llanos de Nisdafe	1854	Agrosystems
	387 Roques de Salmor	661	Cliffs, islets and rocks
	388 Monteverde de Frontera	2447	Monteverde
	389 Western coast of El Hierro	22,996	Tabaibales and cliffs
	390 La Dehesa	2056	Juniper wood and Tabaibales
	391 Bahía de Naos-Hoya de Tacorón	206	Cliffs and islets
Total		31,509	
SAC	ES0000102 Garoé	1124	Monteverde and agrosystems
	ES7020001 Mencáfete	454	Monteverde
	ES7020003 Tibataje	593	Cliffs
	ES7020006 Timijiraque	375	Cliffs
	ES7020026 La Caldereta	18	Monteverde
	ES7020057 Mar de las Calmas	9898	Marina
	ES7020094 Risco de las Playas	966	Cliffs
	ES7020092 Roques de Salmor	4.5	Rocks and islets
	ES7020099 Frontera	8809	Monteverde
Total		22,241.5	
Protected spaces	Frontier Natural Park	12,488	All
	Mencáfete Integral Nature Reserve	469.3	
	Roques de Salmor Integral Nature R	3.5	
	Tibataje Special Nature Reserve	601.6	
	Las Playas Natural Monument	984.8	
	Ventejís Protected Landscape	1143.2	
	Timijiraque Protected Landscape	383	
Total		16,073.4	
International figures	Biosphere Reserve	29,600	
	Geopark	27,800	

Source Biodiversity Data Bank of the Canary Islands

Directive. Its design and methodology were entrusted to Birdlife International, and it was applied in Spain by SEO (Viada 1998). On the island of El Hierro, seven IBAs have been identified based on these criteria (Fig. 3): Macizo de Ventejís, Llanos de Nisdafe, Roques de Salmor, Monteverde de Frontera, western coast of El Hierro, La Dehesa and Bahía de Naos-Hoya de Tacorón.

IBAs to Conserve Forest Birds and Birds of Prey: Macizo de Ventejís and Monteverde de Frontera

The cataloguing of the Macizo de Ventejís as an IBA is due to the presence of an important population of Canary Island Common Chaffinch (subsp. *ombriosa*) and the presence of raptors and various subspecies of endemic passerines, including one of the main populations of Canary Island ravens (90–100 pairs), a species in danger of extinction.

Monteverde de Frontera occupies a wide strip of El Golfo escarpment, hosting the best representation of the Canary Island laurel forest and its stages of degradation on the island of El Hierro. The presence of Bolle's Pigeon (the only population on the island), Canary Island Common Chaffinch (subsp. *ombriosa*) and the Atlantic Canary earned this area its designation as an IBA, in addition to the presence of forest raptors, various subspecies of endemic passerines and a pair of Osprey.

IBAs for Open Space Bird Conservation: Llanos de Nisdafe and La Dehesa

Llanos de Nisdafe is an area characterised by an agro-livestock matrix, in which there are some patches of scrub and heathland. It is a very important area on the island for steppe birds, mainly the Eurasian Stone-curlew and Berthelot's Pipit, and it is a feeding area for various birds of prey and a prime enclave for migrating birds and wintering birds. Its declaration as an IBA is based on all this and, above all, on the presence of Atlantic Canary populations.

La Dehesa is an IBA located to the west of the island of El Hierro dedicated to pastures and agricultural crops, with a Juniper *Juniperus turbinata* woodland and some plantations of foreign pines. This area is of great interest for steppe birds (Eurasian Stone-curlew and Berthelot's Pipit), for migrant birds and for the presence of substantial populations of Common Kestrel.

IBAs to Conserve Seabirds: Roques de Salmor, Bahía de Naos-Hoya de Tacorón and Western Coast of El Hierro

These three IBAs share the same type of ecosystems: coastal cliffs and rocky islets to the north (Roques de Salmor), the south (Bahía de Naos-Hoya de Tacorón) and, above all, on the western coast of the island, extending along the coastal strip and into the sea. These are areas of great interest for the reproduction of Ospreys and seabirds, especially the European Storm Petrel and Band-rumped Storm Petrel, Bulwer's Petrel, Cory's Shearwater, Roseate Tern, Common Tern and Yellow-legged Gull.

The Network of Trails, Paths and Lookouts of El Hierro Constitute the Basic Infrastructure for Ornithological Tourism

The island of El Hierro has two networks of well-consolidated and signposted paths and trails, one was promoted by the Cabildo (Island Government) of El Hierro in accordance with the international standards of the

ERA (European Ramblers Associations) and approved by the FEDME (Spanish Federation of Mountain Sports and Climbing), Hiking of El Hierro, and another dependent on the government of Spain attached to the Natural Trails programme (Traditional paths of El Hierro Natural Trail). Both networks use as their main axis the so-called Camino de la Virgen, 37.3 km long, which links Tamaduste with the Orchilla jetty and has the category of Great Route (GR) (Fig. 4). The Cabildo's network of trails is complemented by eleven Short Route (PR) trails and three local trails (SL), all of which are signposted and have panels with information of interest about the environment and the route itself. This network totals 256 km and covers all the areas of interest for birdwatching on El Hierro. The Nature Trail along the traditional trails of El Hierro is made up of two long-distance trails (Fig. 4): the GR-131 or Camino de la Virgen, which crosses the island through its centre, and a 112 km circular trail that runs around the perimeter of the island, making it possible to visit the different areas and places identified as being of interest for birdwatching.

This network of paths and trails is complemented by a set of nine lookouts (Fig. 4) which, although they were created for scenic and geological purposes, serve as genuine birdwatching points, especially seabirds and birds of prey, as they are located from the coastal cliffs to the peaks of the island. Our proposal is to use these networks of paths and lookouts as a support for birdwatching and the development of ornithological tourism.

5 Birdwatching as a Basis for Bird Conservation: The MacaroAves Project and the Strategy for the Canary Islands

Birdwatching as a tourist activity must contribute to the development of rural areas and the conservation of birds and their habitats, otherwise it is not worth implementing. To achieve this, planning must guide and develop this new economic activity, otherwise it could become a new source of problems in the form of disturbance to endangered species and their habitats. In this sense, since 2016, the Canary Islands have had an Ornithological Tourism Strategy developed within the framework of the MacaroAves project, financed with European Regional Development Fund (ERDF) through the Macaronesia Transnational Cooperation Programme for the period 2007–2013. With a budget of 141,204€, this project was carried out by four partners: SEO/Birdlife in the Canary Islands, SPEA in Madeira, ART in the Azores and Biosfera in Cape Verde. The MacaroAves project aims to make tourism compatible with bird conservation through the implementation of a series of actions: the development of a strategy for ornithological tourism throughout Macaronesia; the identification of the places of

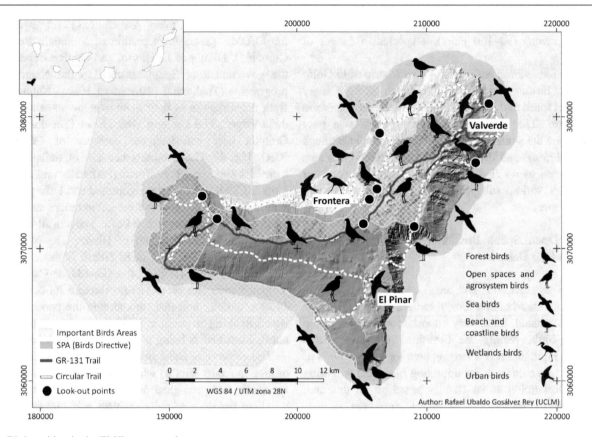

Fig.4 Birdwatching in the El Hierro geopark

greatest interest for ornithological tourism; measures to support the creation of infrastructures; the promotion of this tourism sector as a complementary resource for tourists visiting these archipelagos; the training of local guides and the development of environmental awareness activities.

In the case of the Canary Islands, the main materials produced during the project were the drafting of the Ornithological Tourism Strategy in the Canary Islands, the installation of information panels in demonstration areas in Tenerife and Fuerteventura, the preparation of a guide map of places for birdwatching in the Canary Islands and the organization of a training course in ornithological tourism.

The Ornithological Tourism Strategy of the Canary Islands has identified the potential impacts that poorly planned ornithological tourism could have on birds (SEO/Birdlife 2016). These impacts have been highlighted in multiple works (McFarlane and Boxall 1996; Şekercioğlu 2002; Steven et al. 2011, 2015) and are a concern when implementing birdwatching as a tourism activity.

The challenge for SEO/Birdlife (2016) is to develop ornithological tourism that respects and even enhances natural values based on four fundamental pillars: ensuring the compatibility of activities for the local population and visitors; promoting the conservation of nature (birds, habitats and landscapes); carrying out actions with scientific rigour; and applying a code of ethics and a set of good practices to avoid endangering the species observed. It is worth mentioning that little of this project reached El Hierro, only the creation of a map with fourteen places to observe birds.

In 2018, the *Cabildo* (Island Government) of El Hierro produced an Ecotourism Guide for the Biosphere Reserve and Geopark of the island (Ramos Melo and González del Campo 2018), containing a section dedicated to birdwatching. This guide provides tips for visiting throughout the year and details species and the places where they can be observed. These places are the forest roads of Mencáfete and Jinama, the Llanía spring and the Hoya del Pino recreational area to observe birds of prey and birds typical of the "Monteverde"; Llanos de Nisdafe and San Andrés for birds of steppe and agricultural environments, the reservoirs of Frontera to see migratory waterfowl and the port of La Restinga and the Orchilla lighthouse to observe marine and migratory birds.

6   Conclusion

On the island of El Hierro, little attention has been paid to birdwatching as a tourist activity even though the island is a Biosphere Reserve and Geopark and has numerous nationally and internationally recognised sites. Indeed, the island has bird fauna of great interest due to its endemicity, sites where it is relatively easy to observe the different species and a basic infrastructure of paths and lookouts that allow visitors to travel around the island to conduct this new tourist activity. However, only recently has the Cabildo (Island Government) of El Hierro incorporated bird watching as an activity to be developed in the Biosphere Reserve and Geopark (Ramos Melo and González del Campo 2018).

Ornithological tourism is a solid complement to sun and beach tourism as it favours the development of tourist infrastructures in the interior, is not affected by seasonality and is compatible with short holiday periods or short breaks, an important tool for de-seasonalising tourism. However, it obliges the island's tourism sector to present a differentiated and expert offer for a specific and specialist public. The Geopark and the Biosphere Reserve of El Hierro should make significant efforts in the coming years to support and encourage the development of this new tourist activity, relying on the Ornithological Tourism Strategy for the Canary Islands developed by SEO/Birdlife in the framework of the MacaroAves project. The challenge is, moreover, to do so without deteriorating or endangering these valuable resources, i.e., the birds and their habitats.

References

AVIBASE The World Bird Database. https://avibase.bsc-eoc.org/avibase.jsp?lang=EN

Biggs D, Turpie J, Fabricius C, Spenceley A (2011) The value of avitourism for conservation and job creation—an analysis from South Africa. Conserv Soc 9(1):80–90. https://www.jstor.org/stable/26393127

Canary Islands Natural Inventory Bank (BIOCAN). https://www.biodiversidadcanarias.es/

Carrascal LM, Palomino D (2002) Determinantes de la riqueza de especies en las islas Selvagem y Canarias. Ardeola 49(2):211–221. https://www.ardeola.org/uploads/articles/docs/501.pdf

del Arco Aguilar MJ, Rodríguez Delgado O (2018) Vegetation of the Canary Islands. In: del Arco Aguilar MJ, Rodríguez Delgado O (eds) Vegetation of the Canary Islands. Plant and vegetation, vol 16. Springer, Cham, pp 83–319. https://doi.org/10.1007/978-3-319-77255-4_6

Fernández-Palacios JM, Vera AL, Brito A (2001) Capítulo 17. Ecosistemas. In: Fernández-Palacios JM, Martín Esquivel JL (dirs) Naturaleza de las Islas Canarias. Turquesa Ediciones, Madrid, pp 157–165

Gosálvez Rey RU (2009) El Valle de Alcudia y Sierra Madrona. Paraíso europeo para la observación de aves. Una propuesta de turismo ornitológico. Asociación para el Desarrollo Sostenible del Valle de Alcudia. Ciudad Real. ISBN 978-84-692-9278-5

Hikes of El Hierro. http://senderosdelhierro.com/elhierro-en.php

Jones DN, Buckley R (2001) Birdwatching tourism in Australia. CRC for Sustainable Tourism, Brisbane. ISBN: 1-876685-61-1. https://www.researchgate.net/publication/29458569

Lomolino MV, Riddle BR, Whittaker RJ (2016) Biogeography. Biological biodiversity across space and time. Sinauer Associates, Inc. Sunderland. ISBN 978-1-60535-472-9

López Roig J (2008) Ornithological tourism in the framework of postfordism, a theoretical-conceptual approach. Cuadernos de Turismo 21:85–111. https://revistas.um.es/turismo/article/view/25001

Margalef R (2005) Ecología. Ediciones Omega, S.A. Barcelona. ISBN 978-84-282-0405-7

Martí R, del Moral JC (eds) (2003) Atlas de las aves reproductoras de España. Dirección General de Conservación de la Naturaleza-Sociedad Española de Ornitología, Madrid. ISBN: 8480145501. https://www.miteco.gob.es/es/biodiversidad/temas/inventarios-nacionales/inventario-especies-terrestres/inventario-nacional-de-biodiversidad/ieet_aves_atlas.aspx

Martín A, Lorenzo JA (2001) Aves del Archipiélago Canario. In: Lemus F (ed) La Laguna. ISBN: 978-84-87973-15-4

McFarlane BL, Boxall PC (1996) Participation in wildlife conservation by birdwatchers. Hum Dimens Wildl 1(3):1–14. https://doi.org/10.1080/10871209609359066

Puhakka L, Salo M, Sääksjärvi IE (2011) Bird diversity, birdwatching tourism and conservation in Peru: a geographic analysis. PLoS ONE 6(11):e26786

Ramos Melo JJ, González del Campo P (coord) (2018) Ecotourism guide for the Biosphere Reserve and Geopark of the island of El Hierro. Cabildo Insular de El Hierro. Birding Canarias, SLU. http://www.observatorioelhierro.es/wp-content/uploads/2019/06/GUIA-ECOTURISMO-EL-HIERRO.-21x21guiaecoturismoEN.pdf

Rivera J (2007) Manual con criterios de sostenibilidad para el desarrollo de destinos de aviturismo en Guatemala. Mesa Nacional de Aviturismo-TNC-USAID, Ciudad de Guatemala. https://web.archive.org/web/20140517154300/, http://www.bcienegociosverdes.com/Almacenamiento/Biblioteca/172/2007_Manual_de_criterios_de_sostenibilidad_AVITURISMO_Rivera_J_TNC_GU.pdf

Şekercioğlu Ç (2002) Impacts of birdwatching on human and avian communities. Environ Conserv 29(3):282–289. https://doi.org/10.1017/S0376892902000206

Steven R, Pickering C, Castley JG (2011) A review of the impacts of nature based recreation on birds. J Environ Manage 92(10):2287–2294. https://doi.org/10.1016/j.jenvman.2011.05.005

Steven R, Morrison C, Castley JG (2015) Birdwatching and avitourism: a global review of research into its participant markets, distribution and impacts, highlighting future research priorities to inform sustainable avitourism management. J Sustain Tour 23(8–9):1257–1276. https://doi.org/10.1080/09669582.2014.924955

SEO/BirdLife (2012) Atlas de las aves en invierno en España 2007–2010. Ministerio de Agricultura, Alimentación y Medio Ambiente-SEO/BirdLife. Madrid. ISBN: 978-84-8014-840-5. https://www.miteco.gob.es/es/biodiversidad/temas/inventarios-nacionales/atlas_aves_invierno_tcm30-198034.pdf

SEO/BirdLife (2016) Estrategia de Turismo Ornitológico para la Macaronesia. Canarias. MacaroAves Project. Informe inédito. Santa Cruz de Tenerife. http://www.seo.org/wp-content/uploads/2016/05/Macaroaves_Canarias_estrategia-global_df_baja.pdf

Szczepańska M, Krzyżaniak M, Świerk D, Walerzak M, Urbański P (2014) Birdwatching as a potential factor in the development of tourism and recreation in the region. Barometr Regionalny 12(4):27–38. https://www.researchgate.net/publication/284617268

Traditional Paths of El Hierro Natural Trail. https://www.mapa.gob.es/es/desarrollo-rural/temas/caminos-naturales/caminos-naturales/sector-canario/hierro/default.aspx

Viada C (ed) (1998) Áreas Importantes para las Aves en España. Monografía nº 5. SEO/Birdlife Madrid. ISBN: 84-921901-4-0

Cultural Seascapes in the 'Sea of Calms' and La Restinga Coast

Raquel De la Cruz-Modino, Cristina Piñeiro-Corbeira, Shankar Aswani, Carla González-Cruz, David Domínguez, Paula Ordóñez García, Agustín Santana-Talavera, and José Pascual-Fernández

Abstract

El Hierro has been characterized by the balance between human development and environmental sustainability. The island was historically far from the mass tourism developments dominant on the other Canary Islands. Tourism accommodations in El Hierro are few compared to more developed coastal areas in the Archipelago, and recreational activities are mainly linked to cultural and natural sites and resources. This chapter focuses on La Restinga fishing village and its coasts, where the 'Sea of Calms' and one multiple-use Marine Reserve (MR) are located, both of which became popular over the last decade. The tourist development experience has promoted a new way of looking at the sea and conceptualizing its habitats and populations. In 2014, after the submarine volcano eruption occurred in 2011, we estimated that at least 25,391 dives had been carried out in the diving spots established by the MR and other diving sites close to La Restinga. Despite the difficulties experienced after the volcano eruption, a unique imaginaire has been consolidated, thanks to the image of the island's exclusive nature and iconic elements. In addition, the rapid recovery of the destination is an excellent example of how the tourism system can adapt and incorporate unexpected events such as volcanic eruptions.

Keywords

Tourism · Seascape · Cultural landscape · Submarine volcano eruption

R. De la Cruz-Modino (✉) · C. González-Cruz · D. Domínguez · A. Santana-Talavera · J. Pascual-Fernández
Instituto Universitario de Investigación Social y Turismo.
Universidad de La Laguna, San Cristóbal de La Laguna, Spain
e-mail: rmodino@ull.edu.es

C. González-Cruz
e-mail: cgonzalc@ull.edu.es

D. Domínguez
e-mail: ddomingu@ull.es

A. Santana-Talavera
e-mail: asantana@ull.edu.es

J. Pascual-Fernández
e-mail: jpascual@ull.edu.es

C. Piñeiro-Corbeira
BioCost Research Group, Universidad de La Coruña, A Coruña, Spain
e-mail: c.pcorbeira@udc.es

Present Address:
S. Aswani
Departments of Anthropology and Ichthyology and Fisheries Science, Rhodes University, Grahamstown, South Africa
e-mail: s.aswani@ru.ac.sa

P. Ordóñez García
Green Shark, La Restinga, Spain
e-mail: paupetali@gmail.com

1 Introduction

Marine and coastal seascapes are undoubtedly linked to historical human developments, including a long history of human-environmental interactions across time and space and the accompanying footprints of human activities in the marine environment. Such deep historical interactions have been recorded archaeologically and ethnographically in several locations in the world, including Australia (McNiven 2018), the USA (Erlandson and Jones 2002), Chile (Latorre et al. 2017), and the Canary Islands (Spain) with aboriginal populations (Rodríguez Santana 1996). In the European context, remarkably, the coasts of southern Andalusia (Spain) are home to so-called '*Corrales de Pesca*' (e.g., Florido del Corral 2011, 2014), which exemplify the richness of cultural seascapes that encompass the complexity and diversity of human and environmental relations. They include, but are not limited to, local and territorial knowledge, ichthyological knowledge, and issues of historical-cultural identity related to anthropogenic structures. According to the 2004 European

J. Dóniz-Páez and N. M. Pérez (eds.), *El Hierro Island Global Geopark*,
Geoheritage, Geoparks and Geotourism, https://doi.org/10.1007/978-3-031-07289-5_10

Landscape Convention, landscape concerns natural, cultural, and anthropogenic components as well as their interconnections. Thus, the historical and current forms of human development in different areas and the socio-economic relationships play a role in the cultural foundation of understanding the constitution of cultural seascapes. These undoubtedly include the socio-cultural processes related to leisure time that have conditioned the perception of seascapes (Rodríguez-Darias et al. 2016), incorporating the tourist gaze (Urry and Larsen 2011), and recreational interest and uses (Piñeiro-Corbeira et al. 2020).

In Island contexts, seascape has been regarded as a limiting factor and a part of a more extensive territory in which the sea belongs to the same space as the cultural and economic interconnection of people living on the island (Hau'ofa 1994). Moreover, besides island size and orography, seascape determines the level of openness in which islands may also be exposed to cultural influences from a wider variety of sources (Pungetti 2017), including island colonization, ecological adaptation, and the modern context of economic development such as tourism. Of course, the territory is rarely homogeneous (Crowley 1989), and human adaptations and resource usage may vary, resulting from various cultural seascapes in different island contexts. In the Canary Islands, the linkages with distant cultures date back at some level to the Phoenician-Punics, first, and the Romans later (del Arco Aguilar 2021). It is unclear yet how deep these ancient relationships were. However, the Spaniard conquest (1402–1496) opened the Archipelago to the European trade, converting some islands into a strategic hub for trade between the American, African and European continents (Vieira 2004; Macías 2009), promoting the introduction and development of some agricultural products, especially sugar (Macías 2009). For centuries their ports were an obligatory stop on all routes to the Southern Hemisphere. Santa Cruz de Tenerife and Las Palmas de Gran Canaria ports became essential coal deposits in the Atlantic, particularly during the nineteenth century (Suárez Bosa 2004). Later, the Archipelago went into the health and tourism businesses. Several places on the islands became centers for patients and rich families from Europe (González Lemus and Miranda Bejarano 2002). Marine tourism, including cruises and coastal recreational activities focused on the sea and sun resources, increased and intensified until well into the second half of the twentieth century. Currently, the European commodities demands are still present in large coastal areas and Canarian landscapes. Tourist resorts have occupied the seascape since the 1960s and intensified in the 1980s. The Canary Islands has one of the highest population densities in Spain, and most of the population lives on the coast. Some coastal tourist areas have densities similar to Central-European cities. Most of the 415,287 regulated accommodations in 2017 were concentrated on the four islands with a consolidated tourism sector (Tenerife, Gran Canaria, Fuerteventura, and Lanzarote) (Simancas Cruz and Peñarrubia Zaragoza 2019).

In contrast to these macro processes in some of the islands in the archipelago, El Hierro has been characterized by the balance between human development and environmental sustainability. El Hierro's port was built during the second half of the twentieth century. According to the official tourist accounts, El Hierro has traditionally received fewer tourists than other islands (e.g. 268,405 passengers arrived at the airport in 2019 of whom approximately 133,325 were tourists, according to the Canary Islands Institute of Statistics[1]). The island was not included in the aforementioned trans-oceanic trade routes and was far from the tourist developmental movements on the other islands. Tourism accommodations are few compared to more developed coastal areas in the Archipelago. In 2019 there were only 23,721 travellers staying in tourist establishments in El Hierro.[2] Recreational activities are mainly linked to cultural and natural sites and resources, tangible and intangible ones. On the southern tip of the island, the Marine Reserve (MR) 'Punta de La Restinga-Mar de Las Calmas' (La Restinga Point-Sea of Calms) was declared as *MR with fishing interest* (Jentoft et al. 2012). This chapter focuses on La Restinga and its coasts, where the 'Sea of Calms' and the above-mentioned MR are located (Fig. 1), both of which became the most popular coastal tourist destination in El Hierro over the last decades (De la Cruz Modino et al. 2010). The Sea of Calms takes its name from the, almost year-round, excellent weather conditions in these waters, due to the protection from the dominant winds provided by the island's mountains. The Sea of Calms extends from La Restinga on the east to the Punta Orchilla Lighthouse, the westernmost place in Spain.

2 La Restinga and the Sea of Calms

There was no stable population living in La Restinga until the middle of the twentieth century, although some historians have pointed out that the pier was used as early as 1922 (Acosta Padrón 2003). Initially, some families from the nearby farming village of El Pinar used to go fishing to La Restinga certain times of the year, but it was a temporary endeavour. They used to sleep in caves, between their small boats, or in small huts made with branches and a lichen called 'orchilla' (Galván Tudela 1997). In the 1940s, some

---

[1] Source: Canary Islands Institute of Statistics. Canary Islands Regional Government. 2000–2019. Service Sector.
[2] However, there is an extensive parallel tourist accommodation offer among second residences and holiday-houses, employed by tourists without being officially counted.

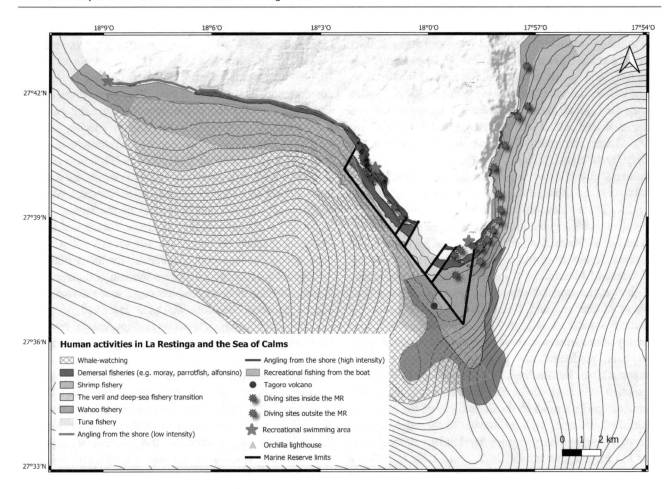

Fig. 1 La Restinga and the Sea of Calms

small-scale fishers from La Gomera (Canary Islands) arrived at La Restinga, and soon established the first permanent settlement. The reasons for these families to move to La Restinga are linked to Francoist repression (Pascual-Fernández et al. 2018) and the conditions in La Gomera (López Felipe 2002). On La Restinga coast and the Sea of Calms (Fig. 1), the good weather allowed fishing throughout the year, and the early permanent fishing settlers could exchange seafood for crops and livestock products with El Pinar and, over time, establish family ties (Pascual-Fernández et al. 2018).

According to the EU definition, fishing activity in La Restinga can be termed as 'small-scale coastal fishing' carried out by fishing vessels with an overall length under 12 m and not using towed gear (Pascual-Fernández et al. 2020), even though many fishers switch their fishing gear from season to season depending on the available species. They employ very selective fishing gear, mainly hand-held artifacts such as canes, hooks and lines, and harpoons and traps for morays or shrimps. Fishers from La Restinga fish mainly within a mile offshore and return to the homeport every day and combine demersal (e.g. blacktail comber, groupers,

alfonsinos) and pelagic fisheries (e.g. skipjack tuna, bluefin tuna), especially for tuna, which is sometimes caught around other islands. Tuna fisheries have been relevant for the entire Canarian fleet and promoted the development of the fishing industry in the southwest coasts of the entire archipelago during the past century. At the end of the 1950s, some tuna canning companies became interested in La Restinga, and began buying local catches and installed an ice factory (De la Cruz Modino 2012). Despite the difficulties that accompany the marketing of artisanal fishery products, the fishermen set up a local cooperative with the support of the government of the island (*Cabildo Insular*) in 1990 (Galván Tudela 1990), which manages most of the products today (Pascual-Fernández et al. 2018). Besides tuna fishing, fishers from La Restinga catch wahoo with harpoons, a legacy of their Gomera origins. These fishers have also developed specific techniques, such as the '*puyón*', which is used to target parrotfish (*Sparisoma cretense*). Fishers snorkel around the rocky bottoms to target individual fish using just a line and a hook, taking advantage of the clear waters.

Migration is part of the history of La Restinga, not just because of its origins. During the second half of the twentieth century, many islanders migrated to other islands and to South America, especially Venezuela. Some of them later returned but, generally, not to the primary sector. Many migrants built apartments, as did others living in other islands. One inmigrant family began promoting scuba diving among the German market, even though the airport was still rudimental and had a small passenger terminal until the 1990s. Later, the scuba-diving businesses expanded, led mainly by foreign families. A generous offer of fresh-fish restaurants consolidated thanks to the local fleet and the involvement of fishing households and women especially (Fig. 2). These changes rendered the community less dependent on fishing, but they maintained the fishing culture and traditions. Since 2000, the population of La Restinga has increased slightly from 443 to 631 inhabitants, 346 men and 285 women.[3] In addition to socio-demographic and territorial change, the tourist development promoted a new way of looking at the sea and conceptualizing its habitats and populations under a new gaze that seeks recreation in marine contemplation. However, these two ways of considering the ecosystem, productive and contemplative, have not been antagonistic in La Restinga. On the contrary, the local population, which maintains critical control over the coastal resources (partly thanks to the declaration of the MR), has established synergies with the service sector, taking advantage of the opportunity opened by tourism development. This has helped give stability to the marketing of different fish products and helped overcome the historical imbalance between men and women in the local population, through the creation of new employment possibilities (De la Cruz Modino and Pascual-Fernández 2005).

3 The Marine Reserve and the Marine Tourism Development

The Sea of Calms is a fragile ecosystem with a high biodiversity, exceptional underwater visibility, and a warm sea surface temperature, making it potentially a great tourism destination, especially for scuba divers (Pascual-Fernández et al. 2015). Some subtropical species are the spotfin burrfish (*Chilomycterus reticulatus*) (Fig. 3f), different rays (e.g., *Mobula tarapacana, Myliobatis aquila, Dasyatis pastinaca, Taeniura grabata*) (Fig. 3b), and sharks such as the angel shark (*Squatina squatina*) and the smalltooth sand tiger shark (*Odontaspis ferox*) (Fig. 3h) occasionally seen in the Sea of Calms (Barría et al. 2018). Also, sea turtles (mainly *Caretta caretta*) (Fig. 3a) and various marine mammals such

as dolphins (e.g., *Tursiops truncatus, Stenella frontalis*) (Fig. 3e), Bryde's whale (*Balaenoptera brydei*), and two beaked whale species, Cuvier's (*Ziphius cavirostris*) and Blainville's (*Mesoplodon densirostris*) (Arranz et al. 2014) are common along these coasts. The narrow underwater shelf, with caves, cliffs, and rocky and sandy habitats, add to the spectacular seascape. Some benthic species, such as groupers (*Epinephelus marginatus, Mycteroperca fusca*) (Fig. 3c), and the common spiny lobster (*Palinurus elephas*) (Fig. 3d), inhabit the rocky and sandy bottoms. La Restinga hosts a statue of a famous grouper called 'Pancho', who died in 2011 and has become a symbol of the natural marine landscape of the area. Before its death, fishers and tourism operators reached an agreement not to fish for groupers, highlighting the importance of Pancho for the destination.

There are currently 15 marine tourism companies in El Hierro, offering various products (e.g., scuba diving, recreational fisheries, leisure). We have identified 34 marine tourism products, which focus on multiple coastal resources. Most companies and activities are located in La Restinga and the MR and work year-round.

The MR was declared in 1996, covering 750 ha, with the support and active involvement of local small-scale fishers and scientists from the University of La Laguna (Tenerife). We consider this MR as an example of a co-governance system, where fishers, scientists, and government cooperated to support the sustainable development of small-scale fisheries and the conservation of marine resources (Pascual-Fernández et al. 2015) (Table 1).

In 2011, in the middle of the 2008–2014 economic crisis in Spain the underwater volcano *Tagoro* erupted giving rise to a novel shallow submarine volcano, seriously affecting marine life, as well as fishing and other marine activities along the coast of El Hierro. In the immediate wake of the volcanic eruption, La Restinga villagers were temporarily evacuated, and professional fishers abandoned all fishing activities from October 2011 to March 2013. However, recreational fishing remained active on parts of the island. Studies from the University of La Laguna showed the area has recovered (Lazzari 2015; Mendoza et al. 2020), and commercial fishers agree. Between 2013 and 2014, fishers came back to their main traditional small-scale fisheries and fishing grounds (Piñeiro-Corbeira et al. 2022). Also, marine tourism activity in the area entirely recovered (Fig. 4). In 2014, we estimated that at least 25,391 dives had been carried out in the diving spots established by the MR and other diving sites close to La Restinga. We also estimated that 2621 diving tourists arrived, spending 609,384 euro in scuba-diving alone during that year (Ordoñez García 2014).

Despite the difficulties experienced after the volcano eruption, a unique imaginaire (Dela Santa and Tiatco 2019) has been consolidated, thanks to the image of the island's exclusive nature and iconic elements, such as Pancho the

---

[3] Source: National Institute of Statistics. Spanish Government. 2020 Demography and population. Register. Population by municipalities. Nomenclator. Santa Cruz of Tenerife.

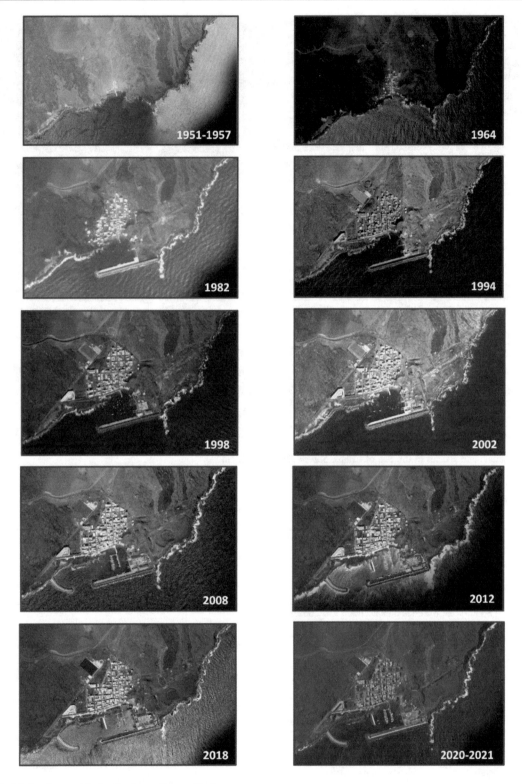

Fig. 2 Territorial development in La Restinga. *Source* Grafcan

grouper and the MR itself. The cultural landscape is created through a seascape that combines two visions: the local view and the tourist gaze. In addition, the rapid recovery of the destination is an excellent example of how the tourism system can adapt and incorporate unexpected events. An example of such capacity is the increase in sports events along the Sea of Calms, such as the *'Travesía a nado Volcán las Calmas'* (Las Calmas Volcano Swimming Tour). This event, unlike the

Fig. 3 Marine habitats and species in the Sea of Calms. *Source* Buceo El Bajón

Table 1 MR characteristics

	Punta de La Restinga-Mar de las Calmas MR			
Responsible	National and autonomous government			
Declaration	1995 (BOC) 1996 (BOE)			
Depth range	0–400 m			
Habitats	Rocky reefs, caves, sandy substrates			
Protection objectives	Small-scale fisheries enhancement and conservation			
Fishers co-management	The *Cofradía* has a strong presence in all governing bodies created or related to the management of the MR			
Forbidden	Anchoring/recreational fishing from boat/spearfishing/scuba diving with propulsion elements/other extractive uses different at described and allowed uses			
Zone classification	**Total MR**	**Maximum restricted area**	**Buffer zone**	**Multiple uses area**
Uses	Professional fishing uses recreational uses	Professional tuna fishery	Professional fisheries	Professional fisheries/scuba diving /angling from the shore/other recreational uses

Source De la Cruz Modino and Pascual-Fernández (2013)

traditional Open Foto-Sub, is available to all participants, not only divers.

As the situation stabilised, the Spanish government proposed the creation of a Marine National Park in the Sea of Calms along the affected coasts. The proposed area included two existing protected areas, one Natura 2000 site and one Special Protection Area (SPA). The proposal focused on the protection of the volcanic cone as well as known marine

Fig. 4 Small-scale fisheries and marine tourism activities in La Restinga and the Sea of Calms. Source: Buceo El Bajón (**a, c, e, g, h**); Manu Machín Quintero (**b, f**); Raquel De la Cruz Modino (**d**)

mammal habitats. Commercial fishers supported the proposal through their *cofradía* (fishers' organization) (Bavinck et al. 2015). However, the recreational fishers opposed it because they felt that a National Park would restrict recreational fishing. Since then, conflicts between recreational fisher groups and small-scale fishers became acrimonious (Pascual-Fernández et al. 2015), although different stakeholders have tried to mediate. Some of them appeared in the local marine governing arena for the first time, such as NGOs that did not take part in the MR creation. The situation, apart from the conflict, is symptomatic of how the cultural landscape is changing with the entry of so many new actors on the territory. The park as projected does not focus on simply protecting fishery resources, as the MR intended, but rather on 'marine resources'. It is also not intended as a fisheries management tool, but rather from the perspective of marine governance, which affects many more public and private agents from the civil society, such as new administrations, not only the traditional fishing responsible. Thus, the protection of the Sea of Calms is not just about the management of fisheries but about marine governance as a whole.

4 Concluding: Cultural Landscape Besides Natural and Fishing Heritage

The landscape, at the intersection between nature and culture, both past and future, as Pungetti and Makhzoumi stated, has a discursive elasticity that encourages its use as a framework for an elastic culture (Pungetti 2017). The Sea of Calms and La Restinga coasts are undoubtedly linked to the fishing tradition, which constitutes the 'sense of place' (Galván Tudela 2003) of this village, which was built in the twentieth century around fishing families that depended on the natural resources of this area and pelagic fish. This natural landscape is particularly fragile due to the small submarine platform surrounding the Island, so fishers are conscious of the need to secure these resources. The MR was built upon this need, but at the same time facilitating the preservation of a submarine and cultural beautiful seascape for visitors, encompassing traditional fishing activities together with marine and scuba diving tourism while securing the natural values of the area (Pascual-Fernández et al. 2018). However, the experience of La Restinga and the Sea of Calms also reveals the active role played by the visitor's gaze, with which locals cohabit. Along with fishers and tourists, new stakeholders and new images (Chuenpagdee et al. 2020) emerge on the Sea of Calms and its resources. As well as environmental emergencies, especially climate change and its effects, including ocean warming and marine ecosystems collapse (Bulleri et al. 2020) may leave their footprint on the cultural landscape. These facts highlight the seascape's dynamism and stakeholders' diversity. The seascape of La Restinga plays an essential role in the provision of cultural services and, despite the difficulties experienced after the submarine volcano eruption, this small-scale fishing village has been consolidated in the tourism market by taking advantage of its remoteness, pristine nature, and exclusivity values.

Acknowledgements This research was supported by the Cajacanarias Foundation and Fundación Bancaria "La Caixa" [grant number 2017REC23], the Ramón Areces Foundation through the XVII Call for

Social Research Grant [grant number CISP17A5887]. This work was also founded by the INTURMAR project [grant number ProID2017010128] and the START-BLUE project [grant number ProID2021010029] integrated into the 'Smart Specialization Strategy of the Canary Islands RIS-3 co-financed by the Operational Program FEDER Canarias 2014–2020' from the Canary Islands Research Agency (ACIISI). RCM, JPF and AST would also like to acknowledge the Macarofood project (Interreg-MAC/2.3d/015), with the support of the European Regional Development Fund. We also would like to thank Mel Cosentino, Manu Machín Quintero and Buceo "El Bajón" for collaborating in the development of this chapter.

References

Acosta Padrón V (2003) El Hierro (1900–1975) Apuntes para su historia. Cabildo insular de El Hierro. Centro de la Cultura Popular Canaria. ISBN: 9788479263799

Arranz P, Borchers DL, de Soto AN, Johnson MP, Cox MJ (2014) A new method to study inshore whale cue distribution from land-based observations. Mar Mamm Sci 30:810–818

Barría C, Colmenero A, Rosario, Del Rosario A, Del Rosario F (2018) Occurrence of the vulnerable smalltooth sand tiger shark, *Odontaspis ferox*, in the Canary Islands, first evidence of philopatry. J Appl Ichthyol 34:684–686

Bavinck M, Jentoft S, Pascual-Fernández JJ, Marciniak B (2015) Interactive coastal governance: the role of pre-modern fisher organizations in improving governability. Ocean Coast Manag 117:52–60

Bulleri F, Sonia B, Sean C, Benedetti Cecchi L, Mark G, Nugues MM, Paul G (2020) Human pressures and the emergence of novel marine ecosystems. Oceanogr Mar Biol Annu Rev 58:441–494

Canary Islands Institute of Statistics. http://www.gobiernodecanarias.org/istac/

Chuenpagdee R, De la Cruz-Modino R, Barragán-Paladines MJ, Glikman JA, Fraga J, Jentoft S, Pascual-Fernández JJ (2020) Governing from images: marine protected areas as case illustrations. J Nat Conserv 3:125756

Crowley J (1989) Landscape ecology, by R.T.T. Forman & M. Godron. John Wiley & Sons, 605 third avenue, New York, NY 10158, USA: Xix 620 pp., figs & tables, 24 × 17 × 3.5 cm, hardbound, US $38.95, 1986. Environ Conserv 16:90–90

del Arco Aguilar MC (2021) De nuevo sobre el descubrimiento y colonización antiguos de Canarias. Reflexiones sobre aspectos teóricos y datos empíricos. Anuario de Estudios Atlánticos 67:1–27

De la Cruz Modino R (2012) Turismo, pesca y gestión de recursos. Aportaciones desde La Restinga y L'Estartit. Ministerio de Educación, Cultura y Deporte, Madrid. ISBN 978-84-8181-489-7

De la Cruz Modino R, Pascual-Fernández JJ (2005) Mujeres, diversificación económica y desarrollo del turismo marino. En torno a la Reserva Marina Punta de la Restinga-Mar de las Calmas (El Hierro-Islas Canarias). In: AKTEA conference: women in fisheries and aquaculture: lessons from the past, current actions and ambitions for the future. La Laguna, Tenerife: Asociación Canaria de Antropología, pp 263–275. ISBN 84-88429-09-6

De la Cruz Modino, R., Pascual Fernández, J.J., Santana Talavera, A., Moreira Gregori, P.E. (2020) Destinos turísticos maduros ante el cambio: reflexiones desde Canarias. Raúl Hernández Martín y Agustín Santana Talavera (coord.), 2010, ISBN 978-84-614-3386-5, págs. 21-48

De la Cruz Modino R, Pascual-Fernández JJ (2013) Marine protected areas in the Canary Islands—improving their governability. In: Bavinck M, Chuenpagdee R, Jentoft S, Kooiman J (eds) Governability

of fisheries and aquaculture, vol 7. MARE Publication Series. Springer, Dordrecht, pp 219–240

Dela Santa E, Tiatco SA (2019) Tourism, heritage and cultural performance: developing a modality of heritage tourism. Tourism Management Perspectives 31:301–309

Erlandson JM, Jones TL (2002) Catalysts to complexity: late holocene cultural developments along the Santa Barbara coast. Perspect California Archaeol 6. ISBN 978-1-931745-08-6

Florido del Corral D (2011) Corrales, una técnica de pesca tradicional en Andalucía. Pescar con arte. Fenicios y romanos en el origen de los aparejos andaluces, pp 65–91

Florido del Corral D (2014) Los corrales de pesca de Cádiz: de derecho señorial a dominio público. Andalucía En La Historia 45:84–89

Galván Tudela JA (1990) "Pescar en grupo" de los azares ambientales a los factores institucionales (La Restinga, El Hierro). Eres. Serie De Antropología 2:39–60

Galván Tudela JA (2003) Sobre las culturas de la mar: prácticas y saberes de los pescadores de La Restinga. El Pajar: Cuaderno De Etnografía Canaria 15:108–117

Galván Tudela JA (1997) La identidad herreña. Centro de la Cultura Popular Canaria, Tenerife. ISBN 84-7926-265-6

González Lemus N, Miranda Bejarano PG (2002) El turismo en la historia de Canarias, Cabildo Insular de Tenerife, La Laguna. ISBN 84-607-6171-1

Hau'ofa E (1994) Our sea of islands. Contemp Pac 6(1):148–161

Jentoft S, Pascual-Fernández JJ, De la Cruz MR, Gonzalez-Ramallal M, Chuenpagdee R (2012) What stakeholders think about marine protected areas: case studies from Spain. Hum Ecol 40:185–197

Latorre C, De Pol-Holz R, Carter C, Santoro CM (2017) I am using archaeological shell middens as a proxy for past local coastal upwelling in northern Chile. Quatern Int 427:128–136

Lazzari N, Hernández JC, Alves F (2015) Resiliencia de las comunidades ícticas de la Reserva Marina del Mar de las Calmas (El Hierro) frente al impacto de la erupción submarina del año 2011. University of La Laguna. Unpublished master thesis

López Felipe JF (2002) La represión franquista en las Islas Canarias. 1936–1950, Editorial Bencomo, La Laguna-Tenerife. ISBN: 978-84-95657-18-3

Macías Hernández, A.M. (2009) Canarias: un espacio transnacional. Reflexiones desde la Historia de la Economía. En A. Galván Tudela (coord.) Migraciones e integración cultural. Las Palmas de Gran Canaria. Academia Canaria de la Historia, pp 95-145.

McNiven IJ (2018) Inhabited Landscapes. The encyclopedia of archaeological sciences. In: López Varela SL (ed). Wiley. ISBN 978-0-470-67461-1

Mendoza JC, Clemente S, Hernández JC (2020) Modeling the role of marine protected areas on the recovery of shallow rocky reef ecosystems after a catastrophic submarine volcanic eruption. Mar Environ Res 155:104877

National Institute of Statistics. https://www.ine.es/

Ordoñez García P (2014) El buceo en el entorno de La Restinga (El Hierro): elementos ambientales, socioeconómicos y de gobernanza. University of La Laguna. Unpublished master thesis

Pascual-Fernández JJ, Chinea-Mederos I, De la Cruz-Modino R (2015) Marine protected areas, small-scale commercial versus recreational fishers: governability challenges in the Canary Islands, Spain. In: Jentoft S, Chuenpagdee R (eds) Interactive governance for small-scale fisheries: global reflections, vol 13. Springer International Publishing, Cham, pp 397–412

Pascual-Fernández JJ, De la Cruz MR, Chuenpagdee R, Jentoft S (2018) Synergy as strategy: learning from La Restinga, Canary Islands. Maritime Stud 17:85–99

Pascual-Fernández JJ, Florido-del-Corral D, De la Cruz-Modino R, Villasante S (2020) Small-scale fisheries in Spain: diversity and challenges. In: Pascual-Fernández JJ, Pita C, Bavinck M (eds) Small-scale fisheries in Europe: status, resilience and governance. Springer International Publishing, Cham, pp 253–281

Piñeiro-Corbeira C, Barreiro R, Olmedo M, De la Cruz-Modino R (2020) Recreational snorkeling activities to enhance seascape enjoyment and environmental education in the Islas Atlánticas de Galicia National Park (Spain). J Environ Manage 272:111065

Piñeiro-Corbeira, C., Barrientos, S., Barreiro, R., Aswani, S., Pascual-Fernández, J.J., De la Cruz-Modino, R. (2022). Can Local Knowledge of Small-Scale Fishers Be Used to Monitor and Assess Changes in Marine Ecosystems in a European Context?. In: Misiune, I., Depellegrin, D., Egarter Vigl, L. (eds) Human-Nature Interactions. Springer, Cham. https://doi.org/10.1007/978-3-031-01980-7_24

Pungetti G (2017) Island landscapes. An expression of European culture, 2016, Routledge

Rodríguez Santana CG (1996) La pesca entre los canarios, guanches y auaritas: las ictiofaunas arqueológicas del Archipiélago Canario. Las Palmas de Gran Canaria: Cabildo Insular

Rodríguez-Darias A, Santana-Talavera A, Díaz-Rodríguez P (2016) Landscape perceptions and social evaluation of heritage-building processes. Environ Policy Gov 26:394–408

Simancas Cruz M, Peñarrubia Zaragoza MP (2019) Analysis of the accommodation density in coastal tourism areas of insular destinations from the perspective of overtourism. Sustainability 11:3031

Suárez Bosa M (2004) The role of the Canary Islands in the Atlantic coal route from the end of the nineteenth century to the beginning of the twentieth century: corporate strategies. Int J Maritime History XVI:95–124

Urry J, Larsen J (2011) The tourist gaze 3.0 SAGE publications. ISBN: 978-1-84920-3777-7

Vieira A (2004) As ilhas atlânticas para uma visão dinâmica da sua história. Anuario De Estudios Atlánticos 50:219–264

Submarine Eruption of El Hierro, Geotourism and Geoparks

William Hernández Ramos, Victor Ortega, Monika Przeor,
Nemesio M. Pérez, and Pedro A. Hernández

Abstract

The year 2011 remained in the memory of the residents of the island of El Hierro (Canary Island, Spain) because of the volcanic episode that originated in its vicinity. From the beginning of the first precursory signs in July 2011, the island's inhabitants reminded that the islands' geological origin is volcanic and, what are the consequent threats of living on them. The eruption, however, has occurred in the marine realm leaving the only threats to the population, strong earthquakes, and diffuse emission of volcanic gases. The Tagoro eruption has not caused any loss of human life, however, its major impact indirectly affected the economy of the residents of the village of La Restinga, in whose vicinity the volcano originated. From a scientific point of view, the eruption has provided an enormous field of observation of the volcanic phenomenon. With the information obtained during the monitoring of the volcano, there is more insight into possible future eruptions. A volcanic product that has never been seen before (Restringolites) was found thanks to this eruption, which is why this volcano was so particular from a petrological point of view. The eruption affected the island's economy, and it also had negative consequences on Herreño tourism. The inhabi-tants of the island, wanting to recover the pre-eruptive economic levels and attract tourists, who, due to the false catastrophic descriptions about the eruption, stopped coming, have taken decisive steps. El Hierro, having peculiar geomorphological and geological characteristics, was the perfect candidate to obtain the Geopark status. In this way, the island of El Hierro, being the Biosphere Reserve since 2000, became also the Geopark since 2014.

Keywords

El Hierro • Historic eruption • Geopark • Restinga • Tagoro

1 Underwater Eruption of El Hierro

The submarine eruption of El Hierro at the end of 2011 is the first of the twenty-first century in the Canary Islands and is the only one in historical period on the island. The eruption of the Tagoro volcano was an eruptive event very well documented by multiple agencies (e.g. Instituto Volcanológico de Canarias (INVOLCAN), Instituto Geográfico Nacional (IGN), Instituto Geológico y Minero de España (IGME), Instituto Español de Oceanografía (IEO), Consejo Superior de Investigaciones Científicas (CSIC), etc.) and managed by Civil Protection with the aim of reducing damage to the population. The eruption originated on the seabed, at a depth of 400 m, in the vicinity of the fishing village of La Restinga (Fig. 1). The economic activity of this area is based on the primary and tertiary sectors, with fishing and diving tourism being the most developed activities.n. However, this volcanic eruption in the vicinity of this area so closely linked to the sea caused direct and indirect consequences on the economy of the residents of La Restinga. The economy of the residents of La Restinga. Even so, thanks to technological and scientific advances, excellent measurements have been obtained before and during the underwater eruption. The locations of the

W. Hernández Ramos (✉) · V. Ortega · N. M. Pérez ·
P. A. Hernández
Volcanological Institute of the Canary Islands (INVOLCAN),
Granadilla de Abona, Spain
e-mail: william.hernandez@involcan.org

V. Ortega
e-mail: victor.ortega@involcan.org

N. M. Pérez
e-mail: nperez@iter.es

P. A. Hernández
e-mail: phdez@iter.es

M. Przeor · N. M. Pérez · P. A. Hernández
Institute of Technology and Renewable Energy (ITER),
Granadilla de Abona, Spain
e-mail: mprzeor@iter.es

J. Dóniz-Páez and N. M. Pérez (eds.), *El Hierro Island Global Geopark*,
Geoheritage, Geoparks and Geotourism, https://doi.org/10.1007/978-3-031-07289-5_11

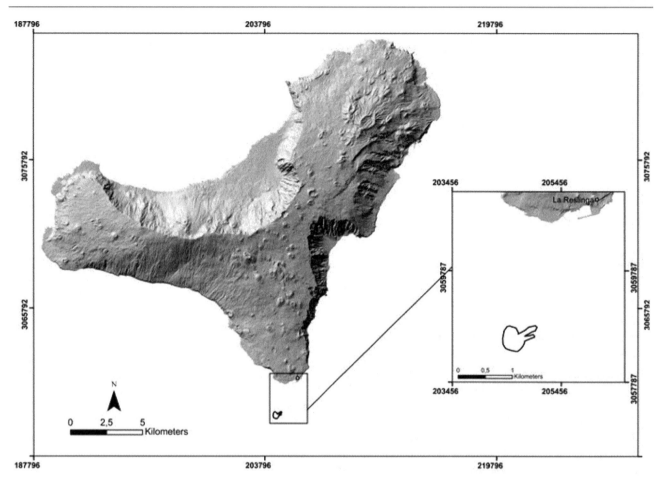

Fig. 1 Location of the Tagoro Volcano south of La Restinga in El Hierro

earthquakes, the deformation of the island, the values of diffuse outgassing of volcanic gases, thermal images and bathymetric representations, allowed us to estimate with great precision the volume of the emitted material. In this sense, although the volcanic hazards linked to the eruption (seismicity, terrain deformation and gas emission) generated risks that affected to a greater or lesser extent the Herreña population, there were no human casualties.

The aim of this chapter is to analyze how the eruption developed from the pre-eruptive to the post-eruptive period and the importance of this volcanic event for the development of the Herreño geopark.

1.1 Pre-Eruptive Period

The pre-eruptive period of the submarine eruption of the island of El Hierro lasted 83 days, starting on 17 July 2011 after the onset of the seismic shock recorded by the geophysical network of the National Geographic Institute (IGN) (Padilla et al. 2013; Pérez-Torrado et al. 2012a, b; Domínguez Cerdeña et al. 2018; Melián et al. 2014; López et al. 2012; Hernández et al. 2013; Pérez et al. 2012, 2014,

2015; Padrón et al. 2013; Ibáñez et al. 2012; Carracedo et al. 2012; Rivera et al. 2013, Sandoval-Velasquez et al. 2021; Rodríguez-Losada et al. 2015; García-Yeguas et al. 2014; Blanco et al. 2015). The seismicity produced had a migratory character (Fig. 2). Throughout the 3-month pre-eruptive period, 12,000 seismic events occurred, migrating in a pattern from the north (El Golfo) to the south (El Julan and the Mar de las Calmas) of the island indicating the movement of magma at depth and the search for the weakest pathways in the crust to emerge (Ibáñez et al. 2012).

Together with seismicity, the IGN recorded ground deformation and an increase in endogenous gas emissions (Pérez et al. 2014; Melián et al. 2014). However, seismicity starred the pre-eruptive period, with earthquakes reaching magnitudes of up to 4.4, creating uncertainty about the place of origin of the new volcano due to its constant spatial migration. On October 8, the largest seismic event in the pre-eruptive stage originated at a depth of 15 km and about 3–4 km from La Restinga. This earthquake was caused by the opening of a hydraulic fracture when magma was injected into the cortical levels (Pérez-Torrado et al. 2012a, b). Some authors claim that the beginning of the eruption was on 10 October, when the harmonic tremor started (Martí

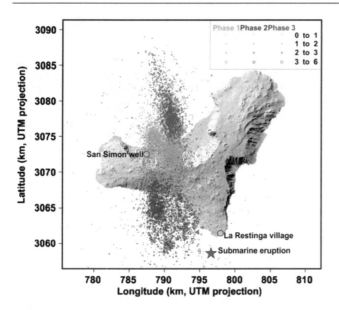

Fig. 2 Location of seismic events recorded by IGN (www.ign.es) up to 5 March 2012 on the island of El Hierro. The colors show different seismic phases described by Ibáñez et al. (2012); the star in red shows the location of the submarine eruption. *Source* Melián et al. (2014)

1.2 Course—Characteristics of the Volcanic Event

During the eruptive stage from October 12, 2011 to March 5, 2012 (206 days) a total of 2500 earthquakes and harmonic tremor (www.ign.es) were recorded. After the start of the eruption, the number of earthquakes was decreasing as the energy accumulated by the magmatic intrusion at depth was released through the eruption. However, on October 20, in the area of El Golfo, the largest number of events and the highest magnitudes were recorded, accompanied by a greater release of pyroclastic material from the eruptive source. These events were interpreted at first as a possible opening of a new eruptive mouth in the north of the island, but no other eruption occurred in the vicinity of it (Perez-Torrado et al. 2012a, b).

The ground deformation experienced as a result of the eruption was of the order of 40 mm in the vertical component and 50 mm in the horizontal component (www.ign.es), with the largest deformation recorded at La Frontera (Fig. 3), with intrusion volumes for the year 2011 estimated at 2.1×10^7 m^3 (Pérez et al. 2014).

The geochemical data also indicated an increase in the diffuse emission of CO-type volcanic gases$_2$ (Fig. 4), H$_2$S, and a significant change in ^{3}He/He^4 ratio and ^{222}Rn activity values (Melián et al. 2014).

In the early stages of the eruption, floating pyroclasts appeared on the sea surface. These volcanic bombs and slags, which were seen for the first time on 15 October 2011, were named Restingolites (Perez-Torrado et al. 2012a, b). It is the first time that pyroclast with white, siliceous cores and black basanitic crust has been studied and documented (Perez-Torrado et al. 2012a, b). Its peculiarity is due to the enormous chemical contrast represented by the product itself. This material emerged from the volcano to the water surface in the first eruptive phases and, contrary to materials with basaltic components, floated in the sea. After the first week of the eruption, the Restingolites stopped appearing on the water surface and on the nearby beaches, and another volcanic product, the "lava baloons" or hollow volcanic bombs, were found inside them (Perez-Torrado et al. 2012a, b). Their collection was complicated because the seawater invaded the interior of the pyroclasts and they lost their buoyancy. This type of volcanic product has been previously documented during submarine eruptions (Clague et al. 2000; Gaspar et al. 2003). In addition to the volcanic materials emitted, discoloration of the water due to gases from the volcanic activity itself was evident, discoloring to different shades of brown, red and green (Fig. 5).

On November 8, 2011 the largest earthquake associated with the eruption of La Restinga was recorded; 2 km from the north coast of the island, with a depth of 21 km and

et al. 2013). However, the first visible evidence of the eruption appeared on October 12 (Pérez et al. 2014). They are the change in color of the seawater, which went from light green to dark brown in the vicinity of La Restinga, as a result of the chemical interaction of seawater with the discharge of hydrothermal fluids at high temperature and magmatic gases. This volcanic manifestation was called *the mancha* (Pérez et al. 2014). The new volcano (Fig. 1) under construction started to be installed on the southern slope of the submarine base of El Hierro.

The main consequences for the population were due to the multiple earthquakes and their high magnitudes. In this sense, on September 23rd the yellow warning level was established in the traffic light which was composed of three colours: red, yellow and green. The most significant measures were the closure of the Los Roquillos tunnel (between the municipalities of La Frontera and El Valverde) and the evacuation of La Restinga. However, in early October 2011 as seismic activity experienced a decrease in frequency and magnitude, the residents of La Restinga returned to their homes. On October 8, 2011 a volcano-tectonic event of higher magnitude (4.4 Ml) occurred and two days later the volcanic tremor started (Ibáñez et al. 2012). The major seismic events caused landslides, rock falls and fear among the population, however, they did not cause major damage among the island's residents. However, the day before the eruption, the residents of La Restinga were evacuated to Valverde as a precaution and to improve the management of the volcanic crisis. volcanic crisis.

Fig. 3 Time series of GPS coordinates for the FRON station from 2011 to the end of 2013, whose reference station is located at GMAS (Maspalomas). Solid red circles represent the elevation determined by the IGN (www. ign.es); first row- horizontal displacement in East–West component; second row-horizontal displacement in North–South component; third row- vertical displacement. *Source* Pérez et al. (2014)

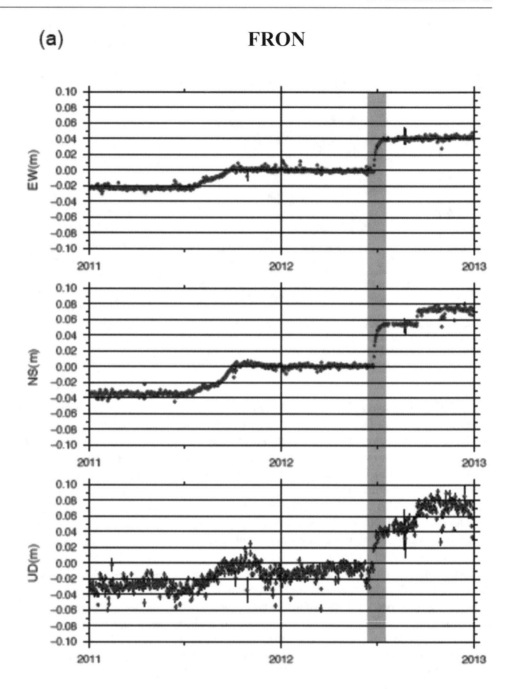

magnitude 4.6Ml (Ibáñez et al. 2012). Since November 24, the volcanic activity had a decrease, noting a drop in seismicity. The newly formed volcano was emitting pillow lavas at depth along with pyroclasts. From 13 February 2012 the tremor was decreasing in amplitude and volcanic activity was markedly reduced, indicating that the end of the eruption was approaching (Ibáñez et al. 2012). After several days without apparent volcanic activity, on March 5, 2012 the IGN reported the end of the eruption.

The IEO made a bathymetric sweep in which the geomorphology of the slope before and after the volcano emerged can be appreciated. The IEO vessel Ramón

Margalef, in the last bathymetric campaign, established that the volcanic cone was 88 m below sea level and the volume of the emitted material was 145×106 m^3 (Perez-Torrado et al. 2012a, b).

The peculiarities of the volcanic event of El Hierro are marked mainly by the volcanic products originated in it as well as by the migratory activity of the seismicity.

During this stage, the main concerns of the authorities were the explosiveness of the eruption as the volcanic cone grew and approached the water surface. The eruption may have been explosive (surtseian activity) and considered hydromagmatic due to the contact of magma with water. In

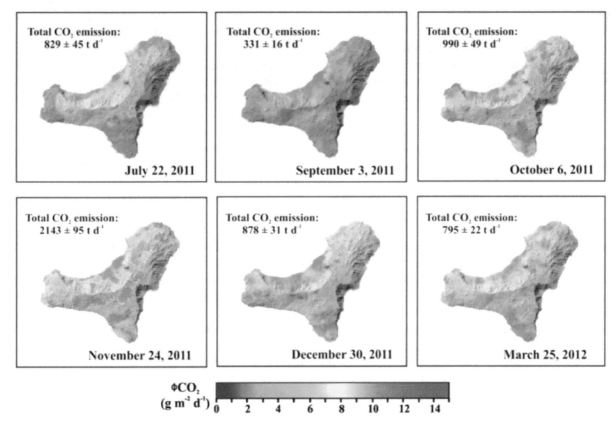

Fig. 4 CO flux maps 2 of El Hierro Island constructed from an average of 100 samples using the sGs method. *Figure* Melián et al. (2014)

Fig. 5 Photograph taken at the source of the submarine volcano by INVOLCAN scientific staff. INVOLCAN SCIENTIFIC STAFF

the case that the eruption had emerged, it could have violently released gases and could have been explosive, so estimating the height of the new volcano with bathymetric studies was of vital importance.

1.3 Post Eruption

The eruption of the Tagoro volcano did not produce any fatalities despite the fact that it was located 1 km off the coast of the village of La Restinga. Seismicity lasted until the following years with high activity and deformation continued until 2013. A decrease in the amount of endogenous gases emitted into the atmosphere was observed. However, this fissural eruption could have had a greater impact on the population and their property if it had had an explosive activity.

2 The Geopark of El Hierro

2.1 Tourism in the Context of the Submarine Eruption

The submarine eruption of the Tagoro volcano had, as has been pointed out, important consequences on the Herreña economy and especially on tourism (Arístegui 2015). During the underwater eruption the main source of the island's economy suffered a severe blow. In December 2011, figures showed that 70% of businesses were closed due to the crisis (https://elpais.com/sociedad/2012/01/19/actualidad/1327003592_267319.html). The figures provided by the Canary Islands Institute of Statistics (ISTAC) prove this. Taking as a reference the year before the eruption, the occupancy rate data for this municipality in October 2010 was 14.5%, in the same month but a year later, when the eruption occurred, the occupancy rate fell to 0.7%. This led to the declaration of a state of social emergency in the three Herreño municipalities and even caused emigration to other islands. Although there were public voices from the locals saying that the volcano was not dangerous, fear and uncertainty set in and caused the unprecedented collapse of this economic activity on the island (https://elpais.com/sociedad/2012/01/19/actualidad/1327003592_267319.html).

This hard time was fostered, in part, by the media's lack of knowledge of this type of event. Parallel to the eruption, there was intense media pressure that caused a false level of alarm to be perceived. In addition, social networks were the focus of hoaxes for visitors, as falsely catastrophic situations were described. These events caused serious damage to the image of El Hierro and caused the tourists present to leave the island (López Moreno 2013). So much so, that the occupancy rate for the year 2011 suffered a slight decline of 1%, compared to the previous year. However, the crisis was even greater the following year, as this percentage fell from 18% in 2011 to 13% in 2012 (ISTAC). If we consider the activity rate for the municipality of El Pinar, to which La Restinga belongs, during the third quarter of 2011 it can be seen that it suffered a decrease of 3.78% compared to the previous quarter of the same year. This is the second largest decline in the activity rate only surpassed by the world crisis COVID-19 (ISTAC).

The underwater eruption caused diving and fishing activities to come to a standstill. The eruption caused sharp temperature gradients, acidification of the water, sulphur and iron concentration (Arístegui 2015). In this context, the Cabildo of El Hierro wanted to reactivate the tourism sector and when the risk traffic light went down to yellow, it wanted to install three viewpoints of geological interest for the observation of the eruption. This was intended to counteract the massive cancellations of bookings in the last months of 2011 (https://www.hosteltur.com/151880_hierro-promociona-fenomeno-volcanico-comom-nuevo-atractivo-turistico-html) through geotourism.

2.2 The Birth of the Geopark of El Hierro

When the eruption ended on March 5, 2012, a new era for tourism on the island began at the same time. The actions that were carried out had as a maxim to take advantage of this fact to relaunch the visit of people. It was then that the idea of turning the island into a geopark was born. This management figure is promoted by the United Nations Educational, Scientific and Cultural Organization (UNESCO) and focuses on a sustained local development that is allowed to manage both cultural and natural heritage, with special emphasis on the volcanic geodiversity of the island (www.elhierrogeoparque.es).

The creation of the Geopark of El Hierro was included in the planning of medium and long term actions to boost the battered economy of El Hierro after the eruption of the submarine eruption. The actions were agreed in several areas such as the Council of Ministers of 28 October 2011, or orders PRE/293636/2011 and IET/460/2012 in which, among other issues, sectoral measures were established to support the promotion of tourism, industrial and business revitalization and the promotion of Information Technology and Communication in El Hierro. The aim of the official bodies was to achieve economic sustainability through the "El Hierro Geopark" brand with initiatives by both public and private business entities (Poch et al. 2015).

The Cabildo of El Hierro and the Ministry of Industry, Energy and Tourism agreed and developed a proposal for the candidacy of El Hierro to join the European Geoparks Network. This proposal was presented in Lesvos (Greece) from 3 to 14 September 2012 and organized by the University of the Aegean with the collaboration of the Global and European Geoparks Network (www.europapress.es/epsocial/responsables/noticia-herro-presienta-portugal-candidatura-entrar-red-europea-geoparques-20120919180113.html).

The process was long and the premises that had to be fulfilled were the following (https://proxecto.xeoparquecaboortegal.gal/es/que-es-que-es-un-geoparque/): That it was a geographically unified territory, with unique geological references and with a certain visual attraction and content for visitors. That there is an entity with management capacity in charge of carrying out an integral strategy related to the conservation of resources, research, education, tourism and economic and social development. That this strategy is in place even before the presentation of the candidacy, with actions to enhance the value of geology and measures to ensure maximum social participation, since geoparks should be built from the bottom up. That there are sufficient economic resources to support the action plan, which can then be carried out directly or with the collaboration of companies and organizations in the territory.

The administrations carried out different improvement works with the aim of responding to the premises to be fulfilled. Among others, the rehabilitation of the Cueva de la Pólvora, the recovery of the Bien de Interés Cultural Fuente de Isora, the opening of the volcanic tube of the Cueva de Guinea, the improvement of the visitor centre Árbol Garoé or the reopening of the Parque Cultural El Julan, with guided routes with ethnographic, historical and geological information. With all this, on 23 September 2014, more than two years after submitting its candidacy, El Hierro was declared a Geopark, thus becoming the first geopark in the Canary Islands. UNESCO took into account several aspects for its inclusion in the European and global network of geoparks, among which stand out that it was a Biosphere Reserve since 2000, that it has seven protected natural areas (Nature Reserves of Mencafete, Roques de Salmor, Tibataje, Protected Landscapes of Ventejís and Timijiraque, Rural Park of Frontera and Natural Monument of Las Playas) that represent 60% of the territory or the experience gained during the crisis of the submarine eruption (www.rtve.es/noticias/20140923/isla-hierro-declarada-geoparque-unesco/1016824.shtml).

2.3 A New Era for El Hierro

One of the main boosts for the economy of El Hierro after the submarine eruption was its declaration as a geopark and the promotion of volcanic geotourism on the island. The geopark area covers the entire emerged island and about 300 km^2 of sea area around El Hierro. Thus, in addition to including the large landslides that are the protagonists in the definition of the geomorphology of El Hierro, the submarine eruption of the Tagoro volcano (Poch et al. 2015) is also included as one of the geozones of the geopark. All these initiatives and the passage of time have made diving regain strength through activities that have been developed for several decades, such as the Open Fotosub underwater photography contest. Likewise, from the public administrations, initiatives were carried out that also had to do in some way with the observation or study of underwater geological

activity. For this reason, the Biosphere, Geology and Geopark Interpretation Centres were built (Poch et al. 2015).

3 Conclusion

The Tagoro eruption between 2011 and 2012 in El Hierro, being submarine in nature, did not cause major damage, although it could have been disastrous if it had originated on land. The only volcanic hazards that took place during the pre- and eruptive period were earthquakes of greater magnitude, however, they did not cause major damage to the population. The deformation of the terrain and the geochemical activity of the gases did not pose a problem for the citizens of the island. More scientific interest was aroused by the volcanic product never seen before-Restingolites, as well as the fact that it was the last volcanic eruption in the Canary Islands for 40 years and the first historical eruption in El Hierro.

On the other hand, the submarine eruption had a significant impact on the Herreña economy. The employment rate fell both for the year as a whole and during the month in which the eruption began. After the end of the eruption there was a turning point that changed the conception of the island and of tourism on El Hierro. In this context, the possibility of turning the island into a Geopark was raised. The attractiveness of its volcanic nature was an important factor, but the fact that the eruption took place below the surface was a major factor, as it added to the attraction it already had for diving activities prior to the eruption. The administrations at both regional and national level were coordinated so that the island ended up being part of the network of geoparks around the world, and this fact was ratified in 2014. In this way El Hierro became the first geopark in the Canary Islands, thus changing the image of the island forever.

References

Arístegui J (2015) Life and death after the submarine eruption of the El Hierro volcano. In: Conference fourth cycle of shared science. ULPGC

Blanco MJ, Fraile-Nuez E, Felpeto A, Santana-Casiano JM, Abella R, Fernández-Salas LM, Vázquez JT (2015) Comment on "Evidence from acoustic imaging for submarine volcanic activity in 2012 off the west coast of El Hierro (Canary Islands, Spain)" by Pérez NM, Somoza L, Hernández PA, González de Vallejo L, León R, Sagiya T, Biain A, González FJ, Medialdea T, Barrancos J, Ibáñez J, Sumino H, Nogami K, Romero C [Bull Volcanol (2014) 76: 882–896]. Bull Volcanol 77(7):1–8

Canary Islands Institute of Statistics. www.gobiernodecanarias.org/istac/

Carracedo JC, Torrado FP, González AR, Soler V, Turiel JLF, Troll VR, Wiesmaier S (2012) The 2011 submarine volcanic eruption in El Hierro (Canary Islands)

Clague D, Moore J, Reynolds J (2000) Formation of submarine flat-topped volcanic cones in Hawai'i. Bull Volcanol 62:214–233. https://doi.org/10.1007/s004450000088

Domínguez Cerdeña I, García-Cañada L, Benito-Saz MA, del Fresno C, Lamolda H, Pereda de Pablo J, Sánchez Sanz C (2018) On the relation between ground surface deformation and seismicity during the 2012–2014 successive magmatic intrusions at El Hierro Island. Tectonophysics 744:422–437. https://doi.org/10.1016/j.tecto.2018.07.019. https://doi.org/10.1016/j.tecto.2018.07.019

El País. https://elpais.com/sociedad/2012/01/19/actualidad/1327003592_267319.html

Europa Press News. www.europapress.es/epsocial/responsables/noticia-herro-presienta-portugal-candidatura-entrar-red-europea-geoparques-20120919180113.html

Europa Press News. www.europapress.es/islas-canarias/noticia-unesco-declara-geoparque-hierro-20140923122145.html

Frontera Town Hall. www.aytofrontera.org/archivo%20historico/plenos/2013/ORDINARIA%2024-04-2013.pdf

García-Yeguas A, Ibáñez JM, Koulakov I, Jakovlev A, Romero-Ruiz MC, Prudencio J (2014) Seismic tomography model reveals mantle magma sources of recent volcanic activity at El Hierro Island (Canary Islands, Spain). Geophys J Int 199(3):1739–1750. https://doi.org/10.1093/gji/ggu339

Gaspar JL, Queiroz G, Pacheco JM, Ferreira T, Wallenstein N, Almeida MH, Coutinho R (2003) Basaltic lava balloons produced during the 1998–2001 Serreta Submarine ridge eruption (Azores). Geophys Monogr-Am Geophys Union 140:205–212

Geopark of El Hierro. www.elhierrogeoparque.es

Geopark Project. https://proxecto.xeoparquecaboortegal.gal/es/que-es-un-geoparque/

Government of the Canary Islands. www.gobiernodecanarias.org/planificacionterritorial/temas/informacion-territorial/enp/hierro/

Hernández PA (2013) The volcanic eruption of El Hierro: the importance of monitoring volcanoes. In: Afonso-Carillo J (ed) El Hierro, birth of a volcano. Proceedings IV scientific week Telesforo Bravo, Institute of Hispanic Studies of the Canary Islands, pp 133–176. ISBN: 978-84-616-5651-6

Hostel Tourism. www.hosteltur.com/138010_hierro-ve-riesgo-erupcion-volcanica-oportunidad-turistica.html

Hostel Turismo. https://www.hosteltur.com/151880_hierro-promociona-fenomeno-volcanico-comom-nuevo-atractivo-turistico-html

Ibáñez JM, De Angelis S, Díaz-Moreno A, Hernández PA, Alguacil G, Posadas A, Pérez NM (2012) Insights into the 2011–2012 submarine eruption off the coast of El Hierro (Canary Islands, Spain) from statistical analyses of earthquake activity. Geophys J Int 191(2):659–670

López Moreno C (2013) Chronicle of an underwater eruption. The seismo-volcanic crisis of El Hierro 2011–2012. Anuario Astronómico del Observatorio de Madrid

López C, Blanco MJ, Abella R, Brenes B, Rodríguez VMC, Casas B, Cerdeña ID, Felpeto A, Villalta MF, de Fresno C, del García O, García-Arias MJ, García-Cañada L, Moreno AG, González-Alonso E, Pérez JG, Iribarren I, López-Díaz R, Luengo-Oroz N, Villasante-Marcos V (2012) Monitoring the volcanic unrest of El Hierro (Canary Islands) before the onset of the 2011–2012 submarine eruption. Geophys Res Lett 39(13). https://doi.org/10.1029/2012GL051846

Martí J, Pinel V, López C, Geyer A, Abella R, Tárrega M, Blanco MJ, Castro A, Rodríguez C (2013) Causes and mechanisms of the 2011–2012 El Hierro (Canary Islands) submarine eruption. J Geophys Res Solid Earth 118(3):823–839. https://doi.org/10.1002/jgrb.50087

Melián G, Hernández PA, Padrón E, Pérez NM, Barrancos J, Padilla G, Dionis S, Rodríguez F, Calvo D, Nolasco D (2014) Spatial and temporal variations of diffuse CO_2 degassing at El Hierro volcanic

system: relation to the 2011–2012 submarine eruption. J Geophys Res Solid Earth 119:6976–6991. https://doi.org/10.1002/2014JB011013

Padilla GD, Hernández PA, Padrón E, Barrancos J, Pérez NM, Melián G, Nolasco D, Dionis S, Rodríguez F, Calvo D, Hernández I (2013) Soil gas radon emissions and volcanic activity at El Hierro (Canary Islands): the 2011–2012 submarine eruption. Geochem Geophys Geosyst 14(2):432–447. https://doi.org/10.1029/2012GC004375

Padrón E, Pérez NM, Hernández PA, Sumino H, Melián G V, Barrancos J, Nagao K (2013) Diffusive helium emissions as a precursory sign of volcanic unrest. Geology 41(5):539–542. https://doi.org/10.1130/G34027

Pérez NM (2015) The 2011–2012 El Hierro submarine volcanic activity: a challenge of geochemical, thermal and acoustic imaging for volcano monitoring. Surtsey Res 13:55–69

Pérez NM, Somoza L, Hernández PA, González de Vallejo L, León R, Sagiya T, Biain A, González FJ, Medialdea T, Barrancos J, Ibáñez J, Sumino H, Nogami K, Romero C (2015) Reply to comment from Blanco et al. on "evidence from acoustic imaging for submarine volcanic activity in 2012 off the west coast of El Hierro (Canary Islands, Spain) by Pérez et al." [Bull. Volcanol. (2014), 76:882–896]. Bull Volcanol. https://doi.org/10.1007/s00445-015-0948-5

Pérez NM, Somoza L, Hernández PA, González de Vallejo L, León R, Sagiya T, Biain A, González FJ, Medialdea T, Barrancos J, Ibáñez J, Sumino H, Nogami K, Romero C (2014) Evidence from acoustic imaging for submarine volcanic activity in 2012 off the west coast of El Hierro (Canary Islands, Spain). Bull Volcanol 76:882. https://doi.org/10.1007/s00445-014-0882-y

Pérez NM, Padilla GD, Padrón E, Hernández PA, Melián GV, Barrancos J, Hernández Í (2012) Precursory diffuse CO_2 and H_2S emission signatures of the 2011–2012 El Hierro submarine eruption, Canary Islands. Geophys Res Lett 39(16)

Pérez-Torrado FJ, Carracedo JC, Rodríguez-González A, Soler V, Troll VR, Wiesmaier S (2012a) The submarine eruption of La Restinga on the island of El Hierro, Canary Islands: October 2011-March 2012. Geol Stud 68(1):5–27. https://doi.org/10.3989/egeol.40918.179

Pérez-Torrado FC, Rodríguez González A, Carracedo JC (2012b) The 2011–12 submarine eruption on El Hierro (Canary Islands): event chronology and crisis management. Geotemas 13:5. ISSN: 1576-5172

Poch J, Montero V Medina Alejandrdo JJ (2015) El Hierro becomes the first geopark in the Canary Islands.

Radio Televisión Española Documentary. www.rtve.es/play/audios/canarias-mediodia/canarias-mediodia-cabildo-hierro-presenta-candidatura-isla-formar-parte-red-europea-geoparques/1540650/

Radio Televisión Española News. www.rtve.es/noticias/20140923/isla-hierro-declarada-geoparque-unesco/1016824.shtml

Rivera J, Lastras G, Canals M, Acosta J, Arrese B, Hermida N, Micallef A, Tello O, Amblas D (2013) Construction of an oceanic island: insights from the El Hierro (Canary Islands) 2011–2012 submarine volcanic eruption. Geology 41(3):355–358. https://doi.org/10.1130/G33863.1

Rodríguez-Losada JA, Eff-Darwich A, Hernández LE, Viñas R, Pérez NM, Hernández PA, Melián G, Martínez-Frías J, Romero-Ruiz C, Coello-Bravo JJ (2015) Petrological and geochemical Highlights in the floating fragments of the October 2011 submarine eruption offshore El Hierro (Canary Islands): relevance of submarine hydrothermal processes. J Afr Earth Sci 102:41–49. https://doi.org/10.1016/j.jafrearsci.2014.11.005

Sandoval-Velasquez A, Rizzo AL, Aiuppa A, Remigi S, Padrón E, Pérez NM, Frezzotti ML (2021) Recycled crustal carbon in the depleted mantle source of El Hierro volcano, Canary Islands. Lithos. https://doi.org/10.1016/j.lithos.2021.106414